Springer Theses

Recognizing Outstanding Ph.D. Research

Aims and Scope

The series "Springer Theses" brings together a selection of the very best Ph.D. theses from around the world and across the physical sciences. Nominated and endorsed by two recognized specialists, each published volume has been selected for its scientific excellence and the high impact of its contents for the pertinent field of research. For greater accessibility to non-specialists, the published versions include an extended introduction, as well as a foreword by the student's supervisor explaining the special relevance of the work for the field. As a whole, the series will provide a valuable resource both for newcomers to the research fields described, and for other scientists seeking detailed background information on special questions. Finally, it provides an accredited documentation of the valuable contributions made by today's younger generation of scientists.

Theses are accepted into the series by invited nomination only and must fulfill all of the following criteria

- They must be written in good English.
- The topic should fall within the confines of Chemistry, Physics, Earth Sciences, Engineering and related interdisciplinary fields such as Materials, Nanoscience, Chemical Engineering, Complex Systems and Biophysics.
- The work reported in the thesis must represent a significant scientific advance.
- If the thesis includes previously published material, permission to reproduce this must be gained from the respective copyright holder.
- They must have been examined and passed during the 12 months prior to nomination.
- Each thesis should include a foreword by the supervisor outlining the significance of its content.
- The theses should have a clearly defined structure including an introduction accessible to scientists not expert in that particular field.

More information about this series at http://www.springer.com/series/8790

Zhiqiang Li

The Source/Drain Engineering of Nanoscale Germanium-based MOS Devices

Doctoral Thesis accepted by
Peking University, Beijing, China

 Springer

Author
Dr. Zhiqiang Li
Institute of Microelectronics
Peking University
Beijing
China

Supervisors
Prof. Xing Zhang
Institute of Microelectronics
Peking University
Beijing
China

Prof. Ru Huang
Institute of Microelectronics
Peking University
Beijing
China

ISSN 2190-5053   ISSN 2190-5061   (electronic)
Springer Theses
ISBN 978-3-662-49681-7   ISBN 978-3-662-49683-1   (eBook)
DOI 10.1007/978-3-662-49683-1

Library of Congress Control Number: 2016934948

Printed on acid-free paper

This Springer imprint is published by Springer Nature
The registered company is Springer-Verlag GmbH Berlin Heidelberg

Parts of this thesis have been published in the following journal articles:

[1] **Zhiqiang Li**, Xia An, Min Li, Quanxin Yun, Meng Lin, Ming Li, Xing Zhang, and Ru Huang, "Low Electron Schottky Barrier Height of NiGe/Ge Achieved by Ion-Implantation after Germanidation Technique," *IEEE Electron Device Lett.*, vol. 33, no. 12, pp. 1687–1689, Dec. 2012.

[2] **Zhiqiang Li**, Xia An, Min Li, Quanxin Yun, Meng Lin, Ming Li, Xing Zhang, and Ru Huang, "Morphology and Electrical Performance Improvement of NiGe/Ge Contact by P and Sb Co-implantation," *IEEE Electron Device Lett.*, vol. 34, no. 5, pp. 596–598, May. 2013.

[3] **Zhiqiang Li**, Xia An, Quanxin Yun, Meng Lin, Min Li, Ming Li, Xing Zhang, and Ru Huang, "Low Specific Contact Resistivity to n-Ge and Well-Behaved Ge n$^+$/p Diode Achieved by Multiple Implantation and Multiple Annealing Technique," *IEEE Electron Device Lett.*, vol. 34, no. 9, pp. 1097–1099, Sep. 2013.

[4] **Zhiqiang Li**, Xia An, Min Li, Quanxin Yun, Meng Lin, Ming Li, Xing Zhang, and Ru Huang, "Study on Schottky Barrier Modulation of NiGe/Ge by Ion-implantation after Germanidation Technique," *The 11th ICSICT*, Xi'an, 2012.

Supervisors' Foreword

With continuous development of integrate circuit in the past 40 years, transistor scaling has been the driving force for technology advancement in the semiconductor industry. However, performance gain becomes more and more difficult as device dimension shrinks into the nanoscale regime, and material innovation is one of the primary enablers for further performance enhancement of CMOS technology. Recently, Ge has attracted much attention owing to its high and symmetric mobilities of electron and hole. Although much improvement has been achieved in Ge-based device, several principal issues still need to be solved for realizing high performance Ge MOSFETs, including the source/drain (S/D) engineering in Ge nMOSFETs.

It is well recognized that one of the critical challenges of nanoscale MOSFET is reducing the impact of parasitic source/drain series resistance, which is more serious in the source/drain of Ge nMOS device, due to the severe Fermi level pinning effect, poor thermal stability of NiGe, and low n-type dopant activation concentration in Ge. Therefore, this Ph.D. thesis by Dr. Zhiqiang Li has focused on the source/drain engineering of Ge nMOSFETs, aiming to reduce the source/drain parasitic resistance. Dr. Li has broad vision and is skilled at identifying the main or controlling factor in daedal situations, and thus significant progress has been made in these fields.

The obvious scientific achievements achieved in this thesis can be divided into several parts. First, implantation after germanide (IAG) technique is applied to modulate the electron SBH of metal/n-Ge contact, and extremely low electron SBH of 0.1 eV is achieved with optimized process parameters. Second, P and Sb co-implantation technique is adopted to improve the thermal stability of NiGe. Both the morphology and electrical performance of NiGe/Ge contact is greatly improved by this technique, and its thermal stability is enhanced at least to 600 °C. Finally, multiple implantation and multiple annealing technique is proposed to reduce the contact resistivity of metal contact to n-Ge, where high electrical activation over 1×10^{20} cm^{-3} and low specific contact resistivity of 3.8×10^{-7} Ω cm^2 is obtained.

These achievements are beneficial for the performance improvement of Ge nMOSFETs.

This thesis has provided effective methods for the aforementioned issues associated with source/drain design of Ge nMOSFETs. Both experiment data and theoretical analysis are accurate and reliable, and the research results have important academic and application value. Thus, this thesis has been highly praised by the eight Ph.D. thesis reviewers and it was rated as an outstanding doctoral dissertation by Peking University. As supervisors of Dr. Li, we are glad to recommend this thesis to readers, particularly those specialized or interested in the related areas.

Beijing   Prof. Xing Zhang
September 2015   Prof. Ru Huang

Abstract

With transistor dimension shrinking into nanoscale regime, it suffer from significant drive current reduction, due to mobility degradation and source/drain parasitic resistance increase. Further development of CMOS technology will strongly depend on the introduction of new material, new process, and new device structure. Germanium (Ge), owing to its high and more symmetric carrier mobility, is considered as a potential channel material for high-performance CMOS devices at 11 nm technology node and beyond. Although Ge pMOSFETs have shown excellent performance, some issues still need to be solved for realizing high performance Ge nMOSFETs, such as poor gate interface quality and large source/drain parasitic resistance. The large source/drain parasitic resistance is an important limiting factor for drive current enhancement of Ge nMOSFETs. Therefore, this thesis focuses on reducing the source/drain parasitic resistance and several feasible solutions have been proposed.

Schottky source/drain is a promising structure for Ge-based device, but the Fermi level pinning effect in metal/Ge contact severely impedes the performance improvement of Ge Schottky Barrier nMOSFET. Therefore, germanium-based Schottky barrier height (SBH) modulation technique is investigated in detail in this thesis. The implantation after germanide (IAG) technique is first proposed to modulate the electron SBH. Phosphorus (P) and arsenic (As) are chosen in the experiment and the influence of the drive-in annealing temperature, implantation dose, and implantation energy on electron SBH modulation is studied. Finally, a record-low electron SBH of 0.10 eV is achieved and good ohmic contact characteristic of NiGe/n-Ge diode is realized, which is promising to promote the performance of Ge-based Schottky barrier nMOSFET.

The self-aligned germanide process is critical to reduce source/drain parasitic resistance for Ge MOSFET, but the poor thermal stability of NiGe puts a roadblock for its application. Therefore, P and Sb co-implantation technique is first proposed in this thesis to improve the thermal stability of NiGe. With this technique, the surface morphology of NiGe film is well improved and the thermal stability of NiGe is at least enhanced to 600 °C. Besides, the electrical characteristic of NiGe/Ge diodes is also improved by this technique. Good ohmic contact characteristic of NiGe/n-Ge diode is also realized and the corresponding contact resistivity is dramatically reduced to 1.2×10^{-6} Ω cm^2. Finally, the P and Sb co-implantation technique is successfully integrated into Ge nMOSFET, and the devices show good output characteristics, which verify that the P and Sb co-implantation technique is favorable for promoting the performance of Ge nMOSFET.

The contact resistance in source/drain comprises the dominant component of parasitic resistance in sub-100 nm MOSFETs, but it is difficult to reduce the contact resistance in Ge-based nMOSFETs, partially due to the poor n-type dopant electrical concentration. Therefore, this thesis addresses the high contact resistance by enhancing the n-type dopant activation in Ge. By optimizing multiple implantation and multiple annealing (MIMA) technique, a high electrical activation over 1×10^{20} cm^{-3} is demonstrated, and the corresponding contact resistivity is reduced to 3.8×10^{-7} Ω cm^2. Meanwhile, the performance of Ge n$^+$/p diodes is also improved by the MIMA technique. The MIMA Ge n$^+$/p diodes exhibit an I_{on}/I_{off} ratio over 10^5, which is much better than the single implanted diode. Finally, the MIMA technique is successfully integrated into Ge nMOSFET, and the saturation current is increased by 6.7 %, which demonstrates that the parasitic resistance is effectively reduced by the MIMA technique.

Keywords Germanium-based MOSFET · Contact resistance · Dopant segregation · Nickel germanide · Dopant activation

Acknowledgments

I am fortunate for the help and support given by so many people along the entire course of my doctoral study. First of all, I would like to express my sincere gratitude to my advisor Profs. Ru Huang and Xing Zhang. Their vast knowledge and strong technical expertise have helped me in getting an excellent background in the semiconductor device technology, identifying the opportunities in this field, and preparing for the challenges. From both the advisors, I not only gained professional knowledge and research method, but also the manner of dealing with people, which is a very valuable asset in my future work and daily life.

I would like to thank Profs. Xia An and Ming Li for their inspiring guidance and constant encouragement throughout this work. I have received a tremendous amount of advice from them, which helped me to successfully resolve many scientific research challenges. During the final writing stage of the thesis, Profs. Xia An and Ming Li guided my writing throughout the day. They not only discussed the framework of the thesis with me, but also carefully weighed every word and corrected punctuation errors many times. Without their help, this thesis would not have been possible.

During my research process, I have also received much help from many seniors in our group. I would like to thank Prof. Yunyi Fu, Prof. Huailin Liao, Dr. Junhua Liu, Dr. Yimao Cai, Dr. Runsheng Wang, and Dr. Le Ye. Meanwhile, special thanks go to senior engineer Xiaoyan Xu for providing advice to process fabrication. In addition, a large portion of this work was achieved in the micro/nano process laboratory of Peking University, and it would not have been possible without the staff's support. I am especially grateful to Prof. Dacheng Zhang and senior engineer Wei Wang for their efficient administrative help.

Next, I want to say thanks to my brothers and sisters in our research group. Special gratitude should be given to my seniors, Yue Guo and Quanxin Yun. Both of them are kindly and helpful, and they gave me much guidance during the experiment. I also thank Min Li, Meng Lin, Pengqiang Liu, Yingxin Qiu, and Zijun Wei for their help during my research. In addition, thanks to our group members for the good times that we shared together; these precious memories will always haunt me in my life.

Finally, I am thankful to my family. Thank you to my father, mother, sister, and cousin for the constant support and endless love. Their encouragements provide me the motivation to move ahead in my career. I also appreciate my girlfriend for her understanding and support. I dedicate this thesis to them.

Contents

1 Introduction 1
  1.1 MOS Device Development Trend in Post-Moore Era 1
  1.2 The Research Status and Challenge of Source/Drain Design 3
    1.2.1 Source/Drain Dopant Activation 3
    1.2.2 Fermi Level Pinning Effect. 4
    1.2.3 NiGe Thermal Stability 6
  1.3 Thesis Organization 6
  References 7

2 Ge-based Schottky Barrier Height Modulation Technology 11
  2.1 The Challenge of Metal Source/Drain in Ge-based Device 11
  2.2 Implantation After Germanidation 14
    2.2.1 The Impact of Drive-in Annealing Temperature. 15
    2.2.2 The Impact of Implantation Energy and Dose 17
    2.2.3 Mechanism Analysis 19
  2.3 Implantation Before Germanidation 20
    2.3.1 P Implantation 21
    2.3.2 Se Implantation. 22
    2.3.3 Comparison of the IAG and IBG Techniques 24
  2.4 Summary 25
  References 25

3 Metal Germanide Technology. 27
  3.1 Thermal Stability Improvement 28
    3.1.1 Experiment Design 28
    3.1.2 Property Analysis 29
    3.1.3 Morphology Improvement of NiGe Film 30
    3.1.4 Thermal Stability Improvement of NiGe Film 31
  3.2 Electrical Performance Improvement of NiGe/Ge Contact 32
    3.2.1 NiGe/Ge Schottky Diodes 32
    3.2.2 NiGe/n-Ge Contact Resistivity 33

3.3   Ge-based Schottky Barrier Source/Drain nMOSFET.   35
      3.3.1   Device Fabrication.   35
      3.3.2   Device Characteristic Analysis   35
3.4   Summary   38
References   39

4   Contact Resistance of Ge Devices   41
4.1   S/D Parasitic Resistance   41
4.2   Multiple Implantation and Multiple Annealing Technique   44
      4.2.1   Experiment Design   44
      4.2.2   N-type Dopant Activation.   45
      4.2.3   Contact Resistivity of Metal/n-Ge Contact   46
      4.2.4   Electrical Characteristics of Ge n^+/p Junction   48
4.3   Ge-based nMOSFETs with MIMA Technique   50
      4.3.1   Device Fabrication.   50
      4.3.2   Device Characteristic Analysis   50
4.4   Summary   53
References   54

5   Conclusions and Prospects   57

Chapter 1
Introduction

With continuous expansion of semiconductor consumption market, integrate circuit (IC) has developed rapidly in the past 40 years. As the foundation of information industry, IC has played an irreplaceable role in commerce, national defense, communication, and daily life. Meanwhile, the information industry based on IC has created remarkable economic benefits, which has exceeded that of traditional industries such as auto, fossil oil, and steel industry. Nowadays, IC is moving toward higher integration density, larger circuit speed, and lower power consumption, and all this will hasten humanity's advance into information age.

1.1 MOS Device Development Trend in Post-Moore Era

Researchers are always trying to reduce the physical dimensions for improving the density, speed, and performance of IC since the first transistor was born in 1948. Moore's law raised by Golden Moore describes that the number of transistors on a chip increases to double at about every 18 months, and the performance of transistor is also improved by scaling the transistor dimensions by 0.7× in every technology node [1, 2]. The biggest manufacture company Intel reported 22 nm node technologies in 2012, and now the 16 and 11 nm node technology are in investigation. The International Technology Roadmap for Semiconductor (ITRS) in Fig. 1.1 predicts the scaling trend of future device gate length, and the gate length is hopeful to scale below 10 nm in the next few years [3]. However, as device gate length scales into nano-regime, short channel effect, high-field effect, and quantum effect have more obvious influence on device performance [4, 5], and further scaling transistor dimensions will be confronted with a series of challenges.

One critical evaluation criterion of MOS device is the intrinsic delay of device, and it can be expressed as follows:

© Springer-Verlag Berlin Heidelberg 2016
Z. Li, *The Source/Drain Engineering of Nanoscale Germanium-based MOS Devices*, Springer Theses, DOI 10.1007/978-3-662-49683-1_1

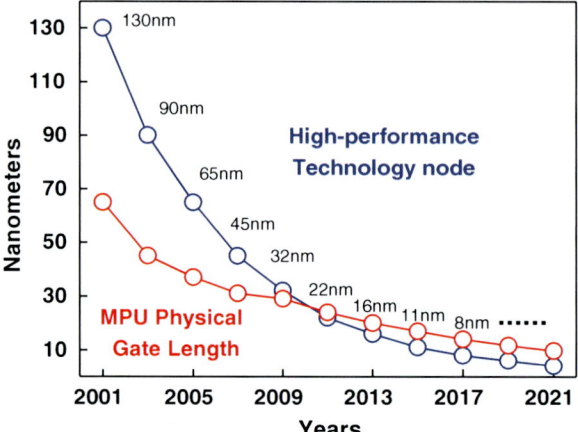

Fig. 1.1 The prediction of physical channel length scaling of future device by ITRS

$$\tau_i = 1/f_i = \left(C_g V_{dd}\right)/I_{on} \tag{1.1}$$

where C_g is gate capacitance, V_{dd} is supply voltage, I_{on} is on-state current. In order to effectively reduce the intrinsic delay of MOS device, the most efficient way is to increase I_{on} as large as possible.

Scaling effective oxide thickness can enhance I_{on} effectively, but now the effective oxide thickness (Eot) has been scaled below 1 nm through adopting high-k dielectric material [6, 7], and further scaling down of Eot will play a limited role for promoting I_{on}. On the other hand, enhancing carrier mobility is another effective method for promoting device driving capability. Strain silicon technology is adopted by Intel in 90 nm technology node for device driving capability improvement [8]. SiGe source/drain is applied in PMOS for increasing carrier mobility through introducing uniaxial compressive stress on channel, and the biaxial tensile stress via a silicon nitride capping layer is similarly introduced in NMOS for drive capability enhancement.

However, when the dimension of device keeps diminishing, the carrier mobility enhancement via strain silicon technology gradually gets saturated. High mobility channel materials such as germanium, As-based III–V semiconductors, graphene, and carbon nanotube are required for further improving carrier injection velocity [9–14]. Table 1.1 gives a summary of the basic characteristic parameters for common high mobility materials [9]. Compared with other high mobility materials, germanium has gained a lot of attention for its potential application as an alternative channel material due to the following reasons. First, the electron and hole mobilities of germanium are 2 times and 4 times larger than that of silicon, respectively, which is very attractive for CMOS integration. Second, the supply voltage will be lower for Ge-based device due to the small bandgap of Ge, which is beneficial for obtaining low power consumption. Third, Ge is compatible with the traditional silicon CMOS

Table 1.1   The basic characteristic parameters of high mobility materials

Property	Material				
	Si	Ge	GaAs	InAs	InSb
Electron mobility (cm²/Vs)	1600	3900	9200	40,000	77,000
Hole mobility (cm²/Vs)	430	1900	400	500	850
Bandgap (eV)	1.12	0.66	1.424	0.36	0.17
Dielectric constant	11.8	16	12.4	14.8	17.7

process for its lower thermal budget, thus it is favorable for integrating with high-k/metal gate structure. In conclusion, germanium is one of the most promising candidates for future advanced CMOS device application, especially in high speed and frequency area.

1.2   The Research Status and Challenge of Source/Drain Design

For over three decades, numerous germanium devices have been demonstrated, such as bulk Ge MOSFETs, germanium-on-insulator MOSFETs and germanium nanowire MOSFETs. Although much improvement has been achieved in Ge-based device, some issues still need to be solved for realizing high performance Ge MOSFETs. The large parasitic resistance in source/drain (S/D) is one of the critical limiting factors for Ge-based device performance improvement, which should be addressed urgently for realizing high performance Ge MOSFETs, especially for Ge nMOSFETs.

1.2.1   Source/Drain Dopant Activation

Ultrashallow junction with high dopant activation is critical to suppress short channel effects and reduce parasitic resistance. However, it is difficult for the source/drain design of Ge nMOS device due to the following reasons. First, the solid solubility of n-type dopants in Ge is relatively low, especially for n-type dopant [15, 16]. The maximum equilibrium solubility of common n-type dopants in Ge (such as P, As, and Sb) is 2×10^{20} cm^{-3} (one magnitude lower than that in Si) [16, 17], and this limits the highest dopant activation concentration in source/drain. Second, the active fraction of n-type dopant in Ge is very low. The maximum achievable electrical concentration of p-type dopant (B) in Ge is 2×10^{20} cm^{-3}, while that of the n-type dopant (P) in Ge is 2×10^{19} cm^{-3}. The much lower n-type dopant electrical concentration in Ge is mainly ascribed to the defects in Ge dominated by vacancies, which act as acceptor state in the bandgap and compensate the donor [18]. Third, the

Table 1.2 The electrical activation achieved by different doping and annealing methods

Doping method	Annealing method	Dopant activation (cm^{-3})
Ion implantation	RTA	P ~ 2×10^{19}/As ~ 8×10^{18} Sb ~ 8×10^{18}/B ~ 1×10^{20} [18]
Ion implantation	Flash anneal	P ~ 6×10^{19} [24]
Ion implantation	Furnace anneal	P ~ 8×10^{18} [26]
Ion implantation	Laser thermal processing	Sb > 1×10^{20} [25]
In situ doping	Thermal	P ~ 1×10^{19} [27]
Gas phase doping	RTA	As ~ 1×10^{19} [22]

diffusivity of dopants in Ge is 2 times higher than that in Ge [19, 20]. Aside from that, the diffusivity of n-type dopants in Ge is larger than that of p-type dopant in Ge [19]. Therefore, it is challenging to obtain well-behaved n$^+$/p Ge diode for application in Ge-based device.

In order to address the above problems, some new ion implantation methods and dopant activation are proposed. Ultra-shallow junctions (X_j < 10 nm) are demonstrated using plasma-immersion ion implantation (PIII) for both n-type and p-type dopants in Ge [21]. Adopting solid source diffusion (SSD) and spin-on dopant (SOD) can effectively reduce the defect introduced during the ion implantation process, which is beneficial for enhancing dopant activation concentration and for controlling dopant diffusion [22, 23]. The electrical activation of n-type dopant achieved by SOD is 7×10^{19} cm^{-3}, and the fabricated junction shows a high I_{ON}/I_{OFF} ratio (10^5–10^6) with an ideality factor of ~1.03 [22]. Meanwhile, high dopant activation concentration can be obtained through advanced annealing methods, such as laser annealing and flash annealing [24, 25]. The electrical activation concentration of n-type dopant using laser annealing is 1×10^{20} cm^{-3} and the I_{ON}/I_{OFF} ratio of the fabricated Ge n$^+$/p junction is over 10^5, but the activation dopant is metastable and the electrical activation may be degenerated when suffering thermal process. Table 1.2 gives the obtained electrical activation with different dopants and annealing methods.

1.2.2 Fermi Level Pinning Effect

As shown in Fig. 1.2, the Fermi level of metal/Ge contact is fixed close to the valance band of Ge, which leads to high electron Schottky barrier height (SBH) of 0.58 eV [28, 29], which is unfavorable for promoting the performance of Ge nMOS device. On the one hand, the high eSBH results in a high specific contact resistivity of metal contact to Ge (>10^{-4} Ω cm^2) [30], which severely restricts the on-state current of Ge nMOS device. Now, the feature size of MOS device is scaled down to 20 nm, and the influence of contact resistance on device performance becomes more notable. On the other hand, the high electron SBH and the low hole SBH make Ge SB nMOSFETs suffer from the ambipolar behavior with insufficient drive current at

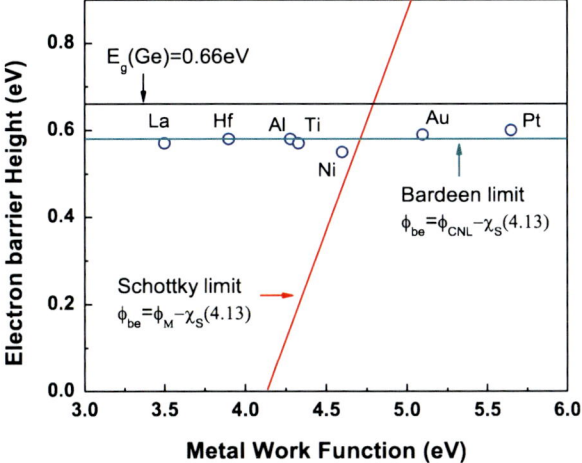

Fig. 1.2   The electron SBHs of different metals contact to Ge

on-state and large leakage current at off-state. Theoretical studies have shown that SB height should be less than 0.1 eV for SB MOSFETs to outperform the doped source/drain devices [31, 32]. Thus, it is urgent to explore a new way to reduce the electron SBH for promoting the performance of Ge nMOSFETs.

In 2004, Connelly first proposed inserting an insulated layer between metal and silicon to suppress metal-induced gap state (MIGS) [33, 34]. As a result, Fermi level pinning at metal/Si Schottky junction was released and Schottky barrier height came to be modulated by tuning metal work function. This method can also be applied in metal/Ge contact, and many insulated layers have been adopted in the eSBH modulation of metal/Ge contact, such as GeO_x [35], Al_2O_3 [36], Y_2O_3 [37], TiO_2 [38], MgO [39], Ge_3N_4 [40], and Si_3N_4 [41]. However, this method can only be used to modulate eSBH. Meanwhile, the inserted insulator also introduces tunneling resistance, which will restrict device drive current.

The S and Se elements have a valence mending adsorbate effect on Ge substrate [42–45], which is beneficial to reduce the interfacial density of metal/Ge contact, and thus helps to alleviate Fermi level pinning effect and reduces eSBH of metal/Ge contact. It has been reported that good ohmic characteristic is obtained using $(NH_4)_2S$ chemical bath treatment [42], and low eSBH (0.13 eV) is achieved using ion implantation of selenium (Se) followed by its segregation at the NiGe/n–Ge interface [43]. However, the S and Se passivation methods are only suitable for eSBH modulation similarly.

Dopant segregation is another SBH modulation technology for metal/Ge contact by introducing a heavily doped Ge layer at the germanide/Ge interface. The hole SBH can be modulated by introducing a heavy boron (B), aluminum (Al), and indium (In) doped layer [46–48]. Similarly, the electron SBH can be modulated by introducing a heavy phosphorus (P), arsenic (As), and antimony (Sb) doped layer

[49, 50]. Since dopant segregation can be used to modulate electron and hole SBH simultaneously with a relatively simple process, it will be particularly discussed in Chap. 2 in this thesis.

1.2.3 NiGe Thermal Stability

Self-aligned germanide process is inevitable in high-performance Ge MOSFETs to maintain low resistance in the source and drain. Among metal germanide materials, NiGe has been considered as the most promising candidate because of its low formation temperature, low resistivity, and feasibility for self-aligned fabrication. However, the poor thermal stability of NiGe is one limiting factor for its application in Ge MOSFET [51, 52]. NiGe film begins to agglomerate at 500 °C, and this leads to the formation of an "island" structure [52, 53]. The rough surface of NiGe will make the film resistance and leakage current increase, severely degrading device performance. In order to enhance the quality of NiGe, different methods have been proposed, such as depositing TiN capping layer and the adoption of nickel alloy by incorporating metal to improve the thermal stability, but these methods only delay the agglomeration to 550 °C [54–57]. Therefore, it is necessary to explore a new way to further enhance the thermal stability and morphology of NiGe film.

1.3   Thesis Organization

Introducing Ge channel material can provide impetus for device performance improvement, however, the large parasitic resistance in source/drain is a significant performance limiting factor for Ge-based device. Therefore, this thesis focuses on the source/drain design and the chapters in this thesis are arranged as follows.

Chapter 1 first presents a brief introduction to the MOSFET scaling challenges and development trends and then, the research status and challenge of source/drain design of Ge MOS device is introduced. Finally, the thesis organization is given.

The eSBH modulation of metal/Ge contact is systematically studied in Chap. 2. First, the IAG method is applied for eSBH modulation of NiGe/Ge contact, and the influence of different annealing and implantation conditions on eSBH modulation is investigated. Second, the IBG method is also used to modulate the eSBH of NiGe/Ge contact with P and Se elements. Finally, the obtained eSBHs of NiGe/Ge contact achieved by IAG and IBG methods are compared.

Chapter 3 focuses on the morphology and thermal stability improvement of NiGe film. First, the reason for the poor thermal stability of NiGe is given and the common methods for NiGe thermal stability enhancement are introduced. Second, co-implantation of P and Sb dopants into NiGe film is proposed to improve the characteristic of NiGe/Ge contact, and its influence on the morphology, thermal stability, and electrical characteristic is discussed. Finally, the P and Sb co-implantation technique is applied in Ge-based SB nMOSFETs, and the corresponding device characteristics are compared.

In Chap. 4, the source/drain contact resistance of Ge MOSFET is studied in detail. First, low specific contact resistivity of metal on n-doped Ge is obtained by optimizing the implantation condition of MIMA technique. Second, the I–V characteristics of Ge n+/p diodes using MIMA technique and single implantation are compared, and the implantation energy is optimized for obtaining a smaller junction depth. Finally, the MIMA technique is introduced in Ge nMOSFETs and the corresponding device characteristics are analyzed.

Chapter 5 summarizes the key contributions of this research and provides suggestions for future research directions.

References

1. Moore GE (1975) Progress in digital integrated electronics. In: IEDM technical digest
2. Moore GE (1998) Cramming more components onto integrated circuits (Reprinted from Electronics, pp 114–117, April 19, 1965). Proc IEEE 86:82–85
3. International technology roadmap for semiconductors. 2013 edn. Available: (http://www.public.itrs.net/)
4. Skotnicki T, Hutchby JA, King TJ, Wong HSP, Boeuf F (2005) The end of CMOS scaling. IEEE Circuits Dev 21:16–26
5. Ieong M, Doris B, Kedzierski J, Rim K, Yang M (2004) Silicon device scaling to the sub-10-nm regime. Science 306:2057–2060
6. Auth C, Allen C, Blattner A, Bergstrom D, Brazier M, Bost M, et al (2012) A 22 nm high performance and low-power CMOS technology featuring fully-depleted tri-gate transistors, self-aligned contacts and high density MIM capacitors. In: 2012 symposium on VLSI technology, VLSIT, pp 131–132
7. Basak S, Nagaraj S, Nahar RK (2013) Simulation and optimization of channel mobility in high-k/metal gate nanoscale MOSFETs. In: Proceedings of international conference on VLSI, communication, advanced devices, signals & systems and networking, VCASAN-2013, pp 231–240
8. Ghani T, Armstrong M, Auth C, Bost M, Charvat P, Glass G, et al (2003) A 90 nm high volume manufacturing logic technology featuring novel 45 nm gate length strained silicon CMOS transistors. In: 2003 IEEE international electron devices meeting, technical digest, pp 978–980
9. Saraswat K (2007) High mobility materials and novel device structures for advanced CMOS technology. Presented at the IEDM short course
10. Saraswat KC, Chui CO, Kim D, Krishnamohan T, Pethe A (2006) High mobility materials and novel device structures for high performance nanoscale MOSFETs. In: 2006 international electron devices meeting, vols 1 and 2, pp 395–398
11. Houssa M, Ye P, Heyns M (2013) High mobility channels. In: High permittivity gate dielectric materials. Springer, Berlin, pp 425–457
12. Schwierz F (2010) Graphene transistors. Nat Nanotechnol 5:487–496
13. Franklin AD, Luisier M, Han S-J, Tulevski G, Breslin CM, Gignac L et al (2012) Sub-10 nm carbon nanotube transistor. Nano Lett 12:758–762
14. Sun Y, Majumdar A, Cheng C-W, Kim Y-H, Rana U, Martin R, et al (2013) Self-aligned III–V MOSFETs: towards a CMOS compatible and manufacturable technology solution. In: 2013 IEEE international electron devices meeting, IEDM, pp 2.7.1–2.7.4

15. Trumbore FA (1960) Solid solubilities of impurity elements in germanium and silicon. Bell Syst Tech J 39:205–233
16. Chui CO, Kulig L, Moran J, Tsai W, Saraswat KC (2005) Germanium n-type shallow junction activation dependences. Appl Phys Lett 87:091909
17. Deal M, Plummer J, Griffin P (2000) Silicon VLSI technology fundamentals, practice and modeling. Prentice Hall, Upper Saddle River
18. Kim J, Bedell SW, Sadana DK (2011) Improved germanium n+/p junction diodes formed by coimplantation of antimony and phosphorus. Appl Phys Lett 98:082112–082112-3
19. Chui CO, Gopalakrishnan K, Griffin PB, Plummer JD, Saraswat KC (2003) Activation and diffusion studies of ion-implanted p and n dopants in germanium. Appl Phys Lett 83:3275–3277
20. Trumbore FA (1960) Solid solubilities of impurity elements in germanium and silicon*. Bell Syst Tech J 39:205–233
21. Thareja G, Chopra S, Adams B, Patil N, Ta Y, Porshnev P, et al (2010) Ultra shallow junctions with high dopant activation and GeO$_2$ interfacial layer for gate dielectric in germanium MOSFETs. In: Device research conference, DRC, 2010, pp 23–24
22. Jamil M, Mantey J, Onyegam EU, Carpenter GD, Tutuc E, Banerjee SK (2011) High-performance Ge nMOSFETs with n(+)-p junctions formed by "spin-on dopant". IEEE Electron Dev Lett 32:1203–1205
23. Chui CO, Kim H, McIntyre PC, Saraswat KC (2003) A germanium NMOSFET process integrating metal gate and improved hi-kappa dielectrics. In: 2003 IEEE international electron devices meeting, technical digest, pp 437–440
24. Wündisch C, Posselt M, Schmidt B, Heera V, Schumann T, Mücklich A et al (2009) Millisecond flash lamp annealing of shallow implanted layers in Ge. Appl Phys Lett 95:252107
25. Thareja G, Liang J, Chopra S, Adams B, Patil N, Cheng SL, et al (2010) High performance germanium N-MOSFET with antimony dopant activation beyond $1 \times 10(20)$ cm(-3). In: 2010 international electron devices meeting—technical digest
26. Kuzum D, Krishnamohan T, Nainani A, Sun Y, Pianetta PA, Wong HSP, et al (2009) Experimental demonstration of high mobility Ge NMOS. In: 2009 IEEE international electron devices meeting, pp 420–423
27. Yu HY, Cheng SL, Griffin PB, Nishi Y, Saraswat KC (2009) Germanium in situ doped epitaxial growth on Si for high-performance n(+)/p-junction diode. IEEE Electron Dev Lett 30:1002–1004
28. Dimoulas A, Tsipas P, Sotiropoulos A, Evangelou EK (2006) Fermi-level pinning and charge neutrality level in germanium. Appl Phys Lett 89:252110–252110-3
29. Nishimura T, Kita K, Toriumi A (2007) Evidence for strong Fermi-level pinning due to metal-induced gap states at metal/germanium interface. Appl Phys Lett 91:123123–123123-3
30. Martens K, Firrincieli A, Rooyackers R, Vincent B, Loo R, Locorotondo S, et al (2010) Record low contact resistivity to n-type Ge for CMOS and memory applications. In: 2010 international electron devices meeting—technical digest
31. Bin Y, Lin JYJ, Gupta S, Roy A, Shurong L, Maszara WP, et al (2012) Low-contact-resistivity nickel germanide contacts on n + Ge with phosphorus/antimony co-doping and Schottky barrier height lowering. In: 2012 international on silicon-germanium technology and device meeting, ISTDM, pp 1–2
32. Xiong SY, King TJ, Bokor J (2005) A comparison study of symmetric ultrathin-body double-gate devices with metal source/drain and doped source/drain. IEEE Trans Electron Dev 52:1859–1867
33. Connelly D, Faulkner C, Grupp DE, Harris JS (2004) A new route to zero-barrier metal source/drain MOSFETs. IEEE Trans Nanotechnol 3:98–104
34. Tersoff J (1984) Schottky barrier heights and the continuum of gap states. Phys Rev Lett 52:465–468
35. Nishimura T, Kita K, Toriumi A (2008) A significant shift of Schottky barrier heights at strongly pinned metal/germanium interface by inserting an ultra-thin insulating film. Appl Phys Exp 1:051406

36. Zhou Y, Ogawa M, Han X, Wang KL (2008) Alleviation of Fermi-level pinning effect on metal/germanium interface by insertion of an ultrathin aluminum oxide. Appl Phys Lett 93:202105–202105-3

37. Li Zhiqiang, An Xia, Yun Quanxin, Lin Meng, Zhang Xing, Huang Ru (2012) Tuning Schottky Barrier height in metal/n-type germanium by inserting an ultrathin yttrium oxide film. ECS Solid State Lett 1(4):Q33–Q34

38. Lin JYJ, Roy AM, Nainani A, Sun Y, Saraswat KC (2011) Increase in current density for metal contacts to n-germanium by inserting TiO_2 interfacial layer to reduce Schottky barrier height. Appl Phys Lett 98

39. Zhou Y, Han W, Yong W, Xiu F, Zou J, Kawakami RK, et al (2010) Investigating the origin of Fermi level pinning in Ge Schottky junctions using epitaxially grown ultrathin MgO films. Appl Phys Lett 96:102103–102103-3

40. Lieten R, Degroote S, Kuijk M, Borghs G (2008) Ohmic contact formation on n-type Ge. Appl Phys Lett 92:022106–022106-3

41. Kobayashi M, Kinoshita A, Saraswat K, Wong HSP, Nishi Y (2009) Fermi level depinning in metal/Ge Schottky junction for metal source/drain Ge metal-oxide-semiconductor field-effect-transistor application. J Appl Phys 105:023702–023702-6

42. Thathachary AV, Bhat KN, Bhat N, Hegde MS (2010) Fermi level depinning at the germanium Schottky interface through sulfur passivation. Appl Phys Lett 96:152108–152108-3

43. Tong Y, Liu B, Lim PSY, Yeo Y-C (2012) Selenium segregation for effective Schottky barrier height reduction in NiGe/n-Ge contacts. IEEE Electron Dev Lett 33:773–775

44. Ikeda K, Yamashita Y, Sugiyama N, Taoka N, Takagi S (2006) Modulation of NiGe/Ge Schottky barrier height by sulfur segregation during Ni germanidation. Appl Phys Lett 88

45. Kaxiras E (1991) Semiconductor-surface restoration by valence-mending adsorbates—application to Si(100)-S and Si(100)-Se. Phys Rev B 43:6824–6827

46. Guo Y, An X, Huang R, Fan CH, Zhang X (2010) Tuning of the Schottky barrier height in NiGe/n-Ge using ion-implantation after germanidation technique. Appl Phys Lett 96

47. Alptekin E, Ozturk MC (2009) Tuning of the nickel silicide Schottky barrier height on p-type silicon by indium implantation. IEEE Electron Dev Lett 30:1272–1274

48. Sinha M, Chor EF, Yeo YC (2008) Tuning the Schottky barrier height of nickel silicide on p-silicon by aluminum segregation. Appl Phys Lett 92

49. Mueller M, Zhao Q, Urban C, Sandow C, Buca D, Lenk S et al (2008) Schottky-barrier height tuning of NiGe/n-Ge contacts using As and P segregation. Mater Sci Eng, B 154:168–171

50. Wong HS, Chan L, Sainudra G, Yeo YC (2007) Sub-0.1-eV effective Schottky-barrier height for NiSi on n-type Si (100) using antimony segregation. IEEE Electron Dev Lett 28:703–705

51. Hsu S-L, Chien C-H, Ming-Jui Y, Huang R-H, Leu C-C, Shen S-W, et al (2005) Study of thermal stability of nickel monogermanide on single- and polycrystalline germanium substrates. Appl Phys Lett 86:251906–251906-3

52. Zhu S, Nakajima A (2005) Annealing temperature dependence on nickel–germanium solid-state reaction. Jpn J Appl Phys 44:L753

53. Lee K, Liew S, Chua S, Chi D, Sun H, Pan X (2004) Formation and morphology evolution of nickel germanides on Ge (100) under rapid thermal annealing. In: MRS proceedings, p C2.4

54. Shiyang Z, Yu MB, Lo GQ, Kwong DL (2007) Enhanced thermal stability of nickel germanide on thin epitaxial germanium by adding an ultrathin titanium layer. Appl Phys Lett 91:051905–051905-3

55. Park K, Lee B, Lee D, Ko D-H, Kwak K, Yang C-W et al (2007) A study on the thermal stabilities of the NiGe and $Ni_{1-x}Ta_xGe$ systems. J Electrochem Soc 154:H557–H560

56. Zhang Y-Y, Oh J, Li S-G, Jung S-Y, Park K-Y, Shin H-S et al (2009) Ni germanide utilizing ytterbium interlayer for high-performance Ge MOSFETs. Electrochem Solid-State Lett 12:H18–H20

57. Zhang Y-Y, Oh J, Li S-G, Jung S-Y, Park K-Y, Lee G-W et al (2010) Improvement of thermal stability of Ni germanide using a Ni–Pt (1 %) alloy on Ge-on-Si substrate for nanoscale Ge MOSFETs. IEEE Trans Nanotechnol 9:258–263

Chapter 2
Ge-based Schottky Barrier Height Modulation Technology

Germanium (Ge) has gained a lot of attention for its potential application as an alternative channel material due to its high and symmetric carrier mobilities. However, due to the low solid solubility and high diffusivity of n-type dopants in Ge, it is very challenging to obtain heavily doped shallow junction. Metal source/ drain (S/D) is considered as a good approach for the S/D engineering, but the performance of Schottky Barrier (SB) MOSFET is still limited by some factors, and one of which is the severe Fermi-level pinning of metal/Ge contact. The high electron Schottky barrier height (SBH) results in SB nMOSFETs suffering from ambipolar behavior with insufficient drive current and large leakage current. And it is very critical to reduce the eSBH of metal/Ge contact for improving Ge-based SB nMOSFET performance. Therefore, the IAG method is first applied in this thesis to reduce the eSBH of NiGe/Ge contact. Besides, the IBG method is also applied to modulate the eSBH for comparison.

2.1 The Challenge of Metal Source/Drain in Ge-based Device

The idea of adopting metal S/D to replace the traditional doped S/D was initially proposed by Nishi in 1966 [1]. Lepselter and Sze first reported the fabricated SB pMOSFET with employing PtSi for the S/D regions [2]. The complementary SB nMOSFET and SB pMOSFET are given in Fig. 2.1. Compared with the common doped S/D structure, the metal S/D has many advantages. First, the low resistivity of metal or germanide is beneficial to reduce the S/D series resistance. Second, the metal S/D has atomically abrupt junctions, and this is favorable for restraining the short channel effect. Finally, its fabrication process is compatible with traditional CMOS process, and the low thermal budget is very beneficial for integrating with high-k/metal gate and strain channel technology.

© Springer-Verlag Berlin Heidelberg 2016
Z. Li, *The Source/Drain Engineering of Nanoscale Germanium-based MOS Devices*, Springer Theses, DOI 10.1007/978-3-662-49683-1_2

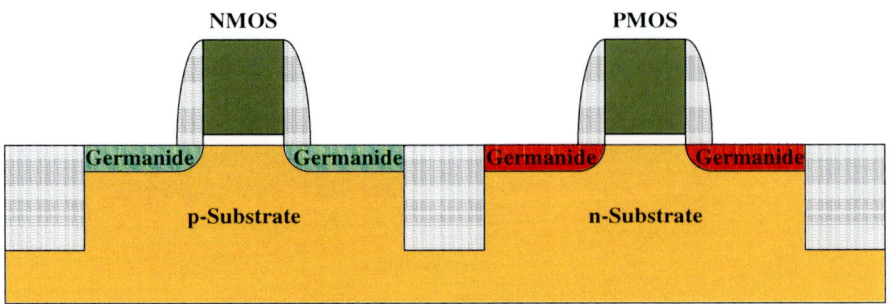

Fig. 2.1 The schematic of complementary-performing SB nMOSFET and SB pMOSFET

However, the high electron SBH is a critical limiting factor for the performance of Ge-based SB nMOSFETs. Since the Fermi level of metal/Ge contact is fixed close to the valance band of Ge, and this leads to a low hole SBH and a high electron SBH. This result is very favorable for Ge SB pMOSFET, because the low hole SBH and high electron SBH are required for obtaining sufficient drive current at the on-state and small leakage current at the off-state, respectively, and the reported Ge SB pMOSFETs have exhibited good electrical characteristics [3–5]. On the contrary, the high electron SBH and the low hole SBH make Ge SB nMOSFETs suffer from the ambipolar behavior with insufficient drive current and large leakage current. And theoretical studies show that SBH should be less than 0.1 eV for SB MOSFETs to outperform the doped source/drain MOSFETs [6, 7]. Therefore, obtaining a sufficient low electron SBH is critical for improving the performance of Ge SB nMOSFETs.

In an ideal case, the electron SBH is determined by the difference between the metal work function and semiconductor electron affinity, and it can be expressed as follows:

$$\emptyset_n = \emptyset_M - \chi_S \tag{2.1}$$

where $\emptyset_n$ is the electron SBH, $\emptyset_M$ is the metal work function, and χ_S is the semiconductor electron affinity. It can be concluded that the electron SBH strongly depends on the metal work function, however, the severe Fermi-level pinning effect makes the electron SBH show weak relation with the metal work function [6, 8–11].

In the case of Fermi-level pinning, the electron SBH can be expressed as follows:

$$\emptyset_n = S(\emptyset_M - \emptyset_s) + (\emptyset_s - \chi_s) \tag{2.2}$$

where $\emptyset_M$ is the metal work function, ϕ_s is the charge neutrality level (CNL), χ_s is the semiconductor electron affinity, and S is the pinning factor. In the Bardeen limit of strong pinning, $S \approx 0$, the barrier height is totally determined by the position of the CNL relative to χ_s. In the Schottky limit of weak pinning, $S \approx 1$, the barrier height is determined by the difference between $\emptyset_M$ and χ_s. In case of metal contact to Ge,

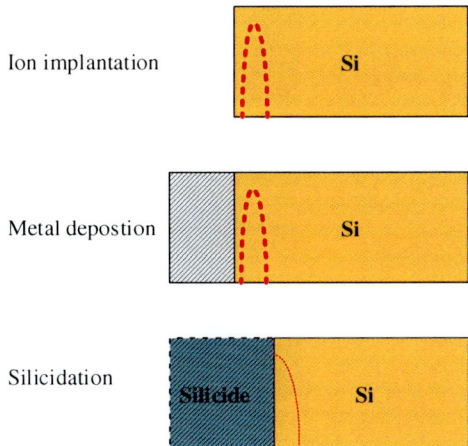

Fig. 2.2 The process flow of IBS technique

the Fermi level is severely pinned with pinning factor S ≈ 0.05–0.02, and the CNL is only about 0.03–0.09 eV above the valence band (VB) [11–13]. As a result of the low-lying CNL and the strong Fermi pinning around this level, the electron SBH is always a high value of about 0.6 eV and the hole SBH is negligible. Therefore, it is quite necessary to explore a new way to reduce the electron SBH.

Dopant segregation technique recently gained significant research attention in reducing the SBH of silicide/silicon contact. The idea behind this technique is that a heavily doped silicon layer formed at the silicide/silicon interface causes a strong conduction/valence band bending near the interface, leading to an effective reduction of the SBHs. R.L. Thornton first achieved the SBH modulation of PtSi/Si contact through dopant segregation method [14]. In this method, the PtSi/Si Schottky diodes are prepared by ion implantation before silicidation (IBS) process [15, 16], and the process flow is given in Fig. 2.2. With scheme IBS, the dopants are expected to be snowplowed and accumulated at the silicide/silicon interface during the silicidation, and the silicide formation will consume the surface Si where implanted dopants are found. For the IBS technique, the process is relatively simple, but ultralow energy ion implantation is usually required and the damage induced from implantation may have a bad effect on the quality of silicide film [17].

Another scheme to incorporate a high concentration of dopant at silicide/silicon interface is the ion implantation after silicidation (IAS) technique as shown in Fig. 2.3 [15, 16]. Compared with IBS method, silicide is first formed in the IAS technique, then dopant is implanted into the silicide film, and the subsequent drive-in anneal process makes dopants diffuse out the silicide film and pile up at the silicide/silicon interface. This technique can avoid the implantation damage and has more process flexibility, which shows great potential for application in CMOS devices.

Similarly, dopant segregation is also applied in Ge-based device for SBH modulation, including implantation before germanidation (IBG) and implantation after

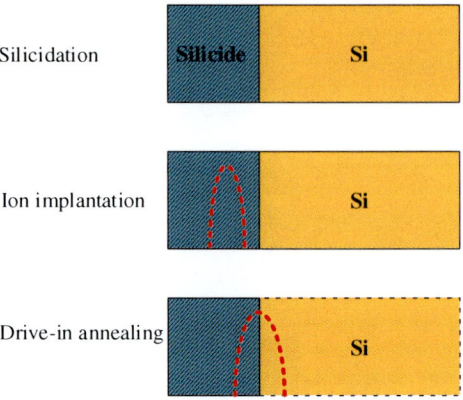

Fig. 2.3 The process flow of IAS technique

germanidation (IAG) to introduce dopant to the germanidation/Ge interface. It has been reported that a low electron SBH of 0.19 eV is achieved by IBG technique with arsenic implantation [18], but the damage introduced by implantation has an obvious influence on the performance of Ge-based device. While for the IAG technique, the implantation damage can be avoided and the process is more flexible. Now, the IAG technique is only applied to modulate the hole SBH of NiGe/Ge contact [19], therefore, this thesis mainly focuses on the electron SBH modulation by IAG technique, and the IBG method is also studied for comparison.

2.2 Implantation After Germanidation

Figure 2.4 gives the process flow of NiGe/Ge Schottky diode with IAG technique. Fabrication of devices were carried out on p-type and n-type Ge(100) substrates with a resistivity of $1–10\,\Omega\,cm$. First, 300 nm SiO_2 film was deposited on the Ge substrates by plasma-enhanced chemical vapor deposition and patterned by the conventional lithography to define the diode areas. After being pretreated with ammonium fluoride pretreatment (AFP) method [20], 20 nm Ni was sputtered and followed by a rapid thermal annealing at 400 °C in nitrogen ambient to form the NiGe film. Then, P^+ and As^+ ions were implanted into NiGe film at the energy range of 33–60 keV. The devices were subjected to drive-in annealing at 350–550 °C for 60 s in nitrogen ambient. Finally, the substrate contact was prepared directly through thermal evaporation of Al onto the Ge backside. The control samples were also fabricated on the n-type and p-type substrates without IAG technique.

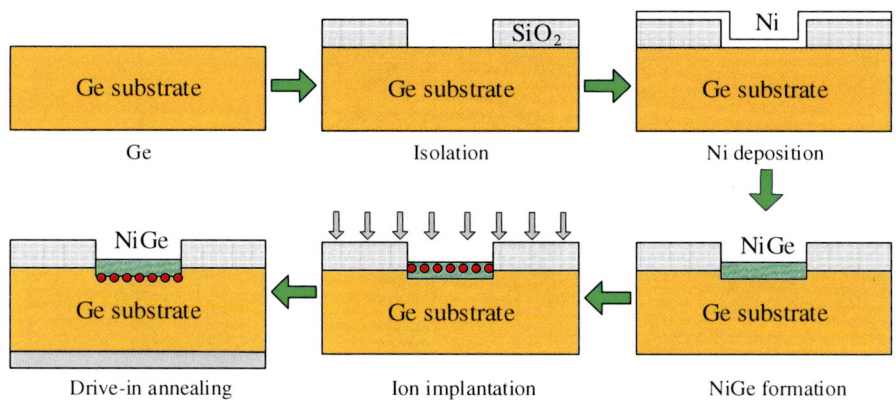

Ge   →   Isolation   →   Ni deposition

Drive-in annealing   ←   Ion implantation   ←   NiGe formation

Fig. 2.4 The process flow of NiGe/Ge Schottky diode with IAG technique

2.2.1   The Impact of Drive-in Annealing Temperature

The I–V characteristics of NiGe/p-Ge (100) Schottky diodes with and without the IAG technique measured at room temperature are given in Fig. 2.5. The control sample without the IAG technique shows well ohmic characteristics, which is because that the Fermi level is pinned in the vicinity of the valance band of Ge. In contrast, the reverse current of the samples with the IAG technique dramatically decreases with the drive-in annealing for the P+ and As+ implantation samples, and the I_{on}/I_{off} ratio of the samples for P+ and As+ implantation annealed at 500 °C is larger than 10^5 and 10^4, suggesting the significant reduction in the electron SBH. While for the drive-in annealing increases to 550 °C, the reverse current of P+ and As+ implantation begins to increase due to the aggregation of NiGe film.

For the NiGe/p-Ge Schottky diodes, the current–voltage (I–V) characteristics can be expressed as follows:

$$I = I_0\left[\exp(qV/nkT) - 1\right], \quad I_0 = AA^*T^2\exp\left(-q\phi_b/kT\right) \tag{2.6}$$

where A is the diode area, A^* is the effective Richardson constant (for p-Ge, $A^* = 48$ Acm^{-2}K^{-2}) [21], T is the temperature in Kelvin, ϕ_b is the hole SBH, and n is the ideality factor. Under forward bias ($V_F > 3kT/q$), the I–V relation can be described as:

$$I = I_0\exp(qV/nkT) \tag{2.7}$$

ϕ_b and n are extracted by linearly fitting of $\ln\left(\frac{J}{AA^*T^2}\right)$ versus $\frac{qV}{kT}$, and Fig. 2.6 illustrates the hole SBHs as a function of the drive-in annealing temperature. It can be seen that the hole SBH gradually increases with the drive-in annealing temperature due to the obvious segregation, its maximum value of 0.56 and 0.52 eV are obtained at 500 °C

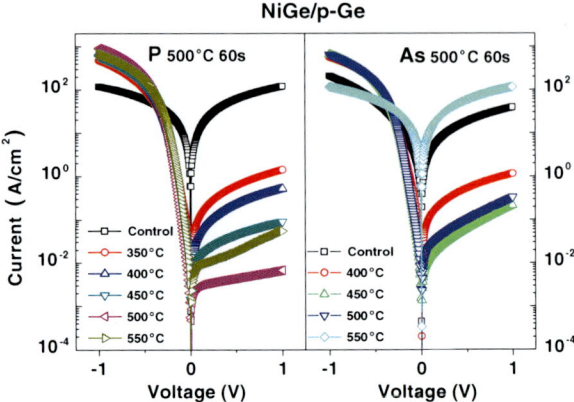

Fig. 2.5 The I–V characteristics of NiGe/p-Ge (100) diodes with different drive-in annealing temperatures

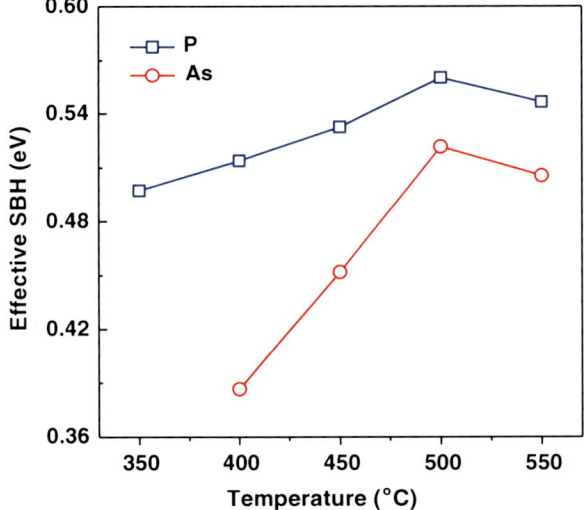

Fig. 2.6 The hole SBHs of NiGe/p-Ge diodes with different drive-in annealing temperatures

for P[+] and As[+] implantation. Since the sum of electron and hole SBHs is approximately the band gap of Ge, the corresponding electron SBH is reduced to 0.10 and 0.14 eV. With the drive-in annealing temperature increasing to 550 °C, the hole eSBH decreases slightly owing to the NiGe film degradation. Therefore, the drive-in annealing temperature should be controlled below 500 °C for the thermal stability restriction of NiGe film.

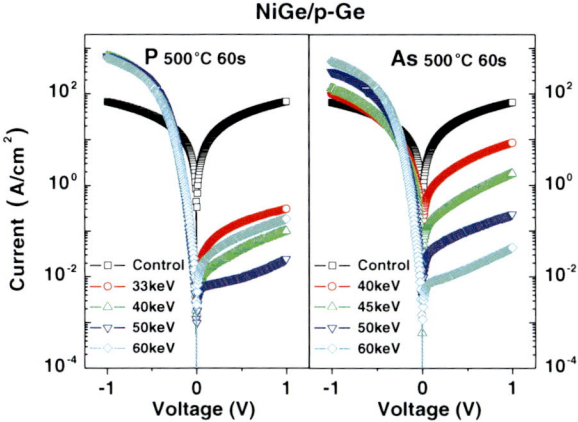

Fig. 2.7 The I–V characteristics of NiGe/p-Ge (100) diodes with different implantation energies

2.2.2 The Impact of Implantation Energy and Dose

Since n-type dopants diffuse quite slowly in NiGe, the ion implantation energy has a direct impact on the effect of dopant segregation and SBH modulation. Therefore, the P+ and As+ implantation energy are studied at the range of 33–60 keV with the fixed drive-in annealing temperature of 500 °C and ion implantation dose of 1×10^{15} cm^{-2}. As shown in Fig. 2.7, the reverse current of NiGe/p-Ge diodes decreases sharply by increasing the implantation energy, and high I_{on}/I_{off} ratios over 10^5 and 10^4 are obtained at 50 and 60 keV for the P+ and As+ implantation samples, indicating the significant SBH modulation. When the ion implantation energy is 60 keV for P+ implantation, the reverse current of NiGe/p-Ge diodes increases again, this is probably ascribed to implantation range over the thickness of NiGe film, which results in a weakened dopant segregation effect.

Accordingly, the extracted hole SBHs shown in Fig. 2.8 increase with the implantation energy, demonstrating its notable influence on SBH modulation. This is because that, the larger implantation energy is beneficial for the implanted dopant to diffuse toward and segregate at the interface, and thus results in more obvious SBH modulation. When the energies are 50 and 60 keV for P+ and As+ implantation, the maximum values of hole SBH are achieved. The implantation energy should be carefully controlled to confine the implanted dopants within the NiGe film, otherwise n+/p junction will be formed.

Similarly, the impact of implantation dose on SBH modulation is also studied. The implantation energy and drive-in annealing temperature are fixed at optimum value for P+ and As+ implantation, respectively, and the implantation dose is set at the range from 1×10^{13} to 2×10^{15} cm^{-2}. The I–V characteristics of NiGe/p-Ge (100) Schottky diodes as a function of different implantation dose are given in Fig. 2.9. It can be seen that the reverse current of NiGe/p-Ge diodes decreases sharply by

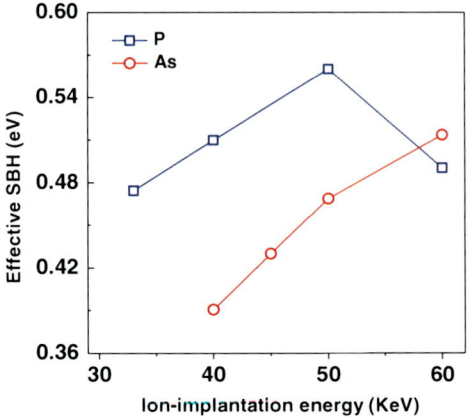

Fig. 2.8 The hole SBH of NiGe/p-Ge diodes with different ion implantation energies

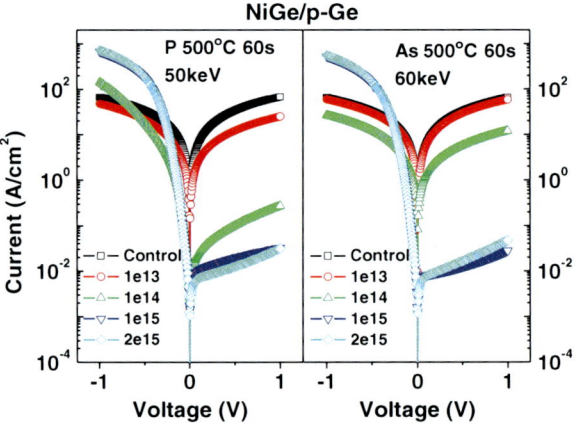

Fig. 2.9 The I–V characteristics of NiGe/p-Ge (100) diodes with different implantation doses

increasing the P^+ and As^+ implantation dose, suggesting the significant increase in the hole SBH. When the implantation dose is increased over 1×10^{15} cm^{-2}, the reverse current of NiGe/p-Ge diodes no longer decreases, which is possibly ascribed to the restriction of dopant activation in Ge [22].

The hole SBHs of NiGe/p-Ge diodes with different implantation doses are given in Fig. 2.10. It can be seen that, for both P^+ and As^+ implantation, below the implantation dose of 1×10^{14} cm^{-2}, the hole SBH is lower than 0.40 eV, and then it drastically elevates over 0.5 eV at 1×10^{15} cm^{-2}. By further increasing the implantation dose to 2×10^{15} cm^{-2}, the hole SBH shows a slight increase, which is possibly due to the saturation of dopant activation at high implantation dose. Based on the above results, the optimum implantation doses for both P^+ and As^+ is 1×10^{15} cm^{-2}.

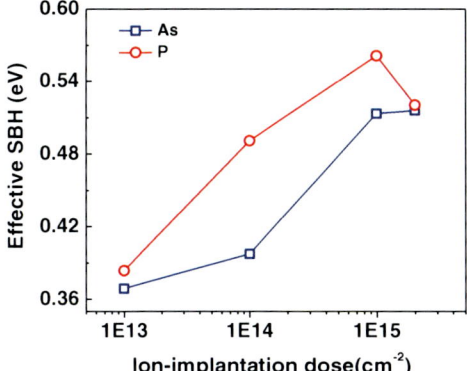

Fig. 2.10 The hole SBH of NiGe/p-Ge diodes with different ion implantation doses

2.2.3   *Mechanism Analysis*

The mechanism for SBH modulation is attributed to the formed dipoles at the NiGe/Ge interface resulted by dopant segregation. Figure 2.11 gives the secondary ion mass spectroscopy (SIMS) analysis of the P+ samples with the drive-in annealing at 350 and 500 °C, phosphorus concentration peaks are observed near the NiGe/Ge interface, and the P+ sample annealing at 500 °C exhibits a higher phosphorus concentration peak due to the higher annealing temperature. With the segregated P atoms at the interface occupying the substitutional sites and being positively charged, dipoles across the interface are formed. These dipoles then lead to the Ge energy band bending downward, which leads to a smaller effective hole SBH, and thus the corresponding electron SBH is reduced [23].

In addition, the NiGe/n-Ge diodes were fabricated with optimized process parameters to validate the effectiveness of electron SBH modulation. As it is shown in Fig. 2.12, the control sample without the IAG technique shows typical rectifying characteristic due to the high electron SBH resulted by Fermi-level pinning effect, whereas well ohmic characteristics are obtained for the IAG samples with drive-in annealing temperature over 350 and 450 °C for P+ and As+ implantation, respectively. Compared with As+, P+ has a wider process window between 350 and 500 °C for obtaining optimized SBH, and this is because that the implanted P+ is easier to diffuse toward and segregate at the NiGe/Ge interface with smaller atomic radius. In conclusion, an effective SBH modulation of NiGe/Ge contact has been experimentally demonstrated by the P+ and As+ IAG technique, and this is very attractive for improving the performance of Ge-based devices.

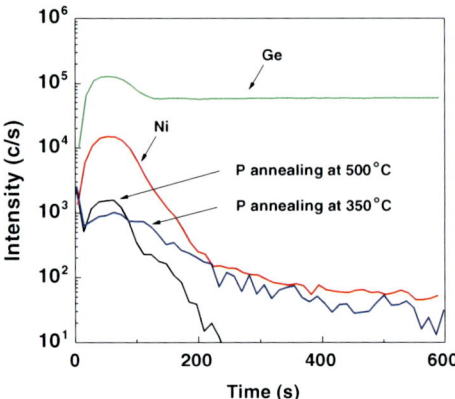

Fig. 2.11 Phosphorus concentration profile measured by SIMS after the drive-in annealing at 350 °C and 500 °C

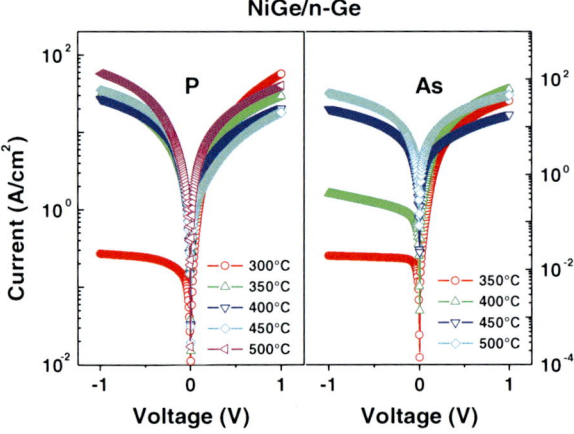

Fig. 2.12 The I–V characteristics of NiGe/n-Ge (100) diodes with IAG technique

2.3 Implantation Before Germanidation

The implantation before germanidation (IBG) is also applied to SBH modulation with phosphorus (P) and selenium (Se) implantation, and the influence of annealing temperature on SBH modulation is investigated. Figure 2.13 gives the process flow of NiGe/Ge Schottky diode with IBG technique.

The n-type Ge(100) substrates with a resistivity of 1–10 Ω cm were used for device fabrication. 300 nm SiO_2 film was deposited on the Ge substrates by plasma-enhanced chemical vapor deposition and patterned by the conventional lithography to define the diode areas, and the SiO_2 film was etched down to 150 Å as implantation

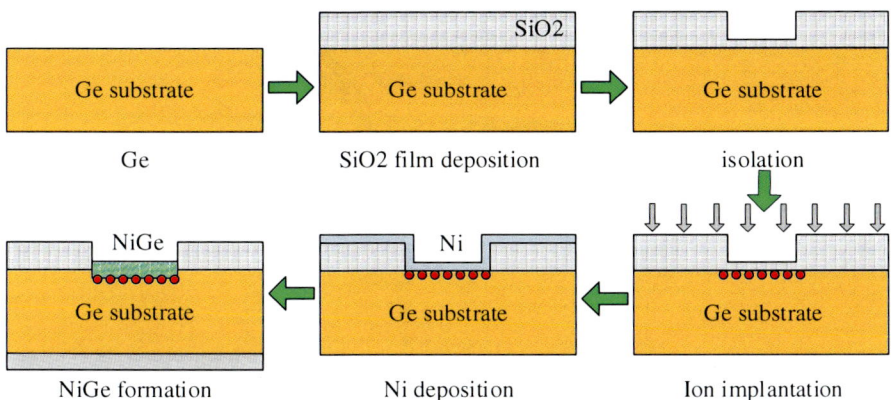

Fig. 2.13 The process flow of NiGe/Ge Schottky diode with IBG technique

mask. Then, P+ and Se+ ions were implanted into NiGe film with 24 keV and 42 keV, respectively. After that, the 150 Å implantation mask was removed by buffered oxide etch (BOE). The wafers were pretreated with AFP method, and 30 nm Ni was sputtered and followed by a rapid thermal annealing at 350–500 °C in nitrogen ambient to form the NiGe film. Finally, the substrate contact was prepared directly through thermal evaporation of Al onto the Ge backside.

2.3.1   P Implantation

Figure 2.14 gives the I–V characteristics of NiGe/n-Ge (100) Schottky diodes. As it is shown, the control sample without the IBG technique shows typical rectifying characteristics, which is consistent with the high electron SBH resulted by Fermi-level pinning effect. With P+ implantation and annealing at 350 °C, dopants pile up at the formed NiGe/Ge interface, and the electron SBH is reduced, which leads to the reverse current increasing by 2–3 order of magnitudes. When the annealing temperature is increased to 400 °C, a decrease of reverse current is observed, which is possibly owing to the reduced dopant concentration at the contact interface with higher annealing temperature.

Figure 2.15 gives the rectification ratio and electron SBH of NiGe/n-Ge diodes as a function of annealing temperature. The electron SBH of control sample is 0.52 eV, which is consistent with the published data [19]. With annealing at 350 °C, the rectification ratio of NiGe/n-Ge diodes is 1.3, indicating well ohmic characteristics. However, the rectification ratio and the electron SBH begin to increase when the annealing temperature increases to 400 °C, owing to dopant diffusion toward the substrate at higher annealing temperature. Therefore, the annealing temperature should be controlled below 400 °C to obtain good SBH modulation effects with P+ implantation.

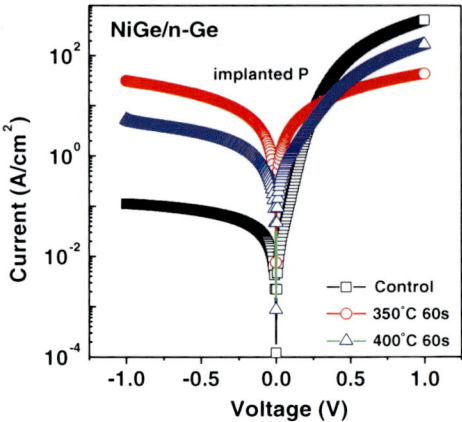

Fig. 2.14 The I–V characteristics of NiGe/n-Ge (100) diodes with different annealing conditions

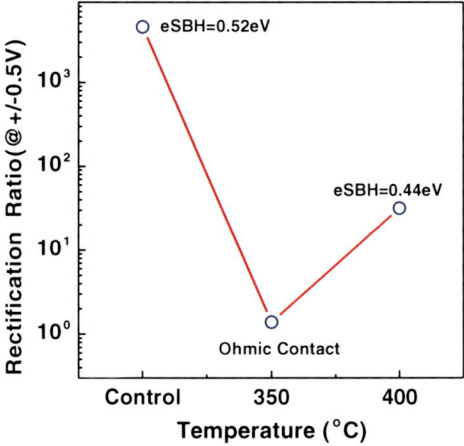

Fig. 2.15 The rectification ratio and electron SBH of NiGe/n-Ge diodes

2.3.2   Se Implantation

The effect of Se on SBH modulation of NiGe/Ge contact is also investigated in this thesis. Figure 2.16 gives the I–V characteristics of NiGe/n-Ge (100) Schottky diodes with Se implantation as a function of annealing temperature. It can be seen that the reverse current of NiGe/n-Ge (100) Schottky diodes gradually increases by increasing the annealing temperature (from 350 to 500 °C), and well ohmic characteristic with rectification ratio close to 1 is obtained when the annealing temperature is 450 °C, indicating the effective SBH modulation of the NiGe/Ge contact. But the current of the Se implanted sample is smaller than that of the P implanted sample as

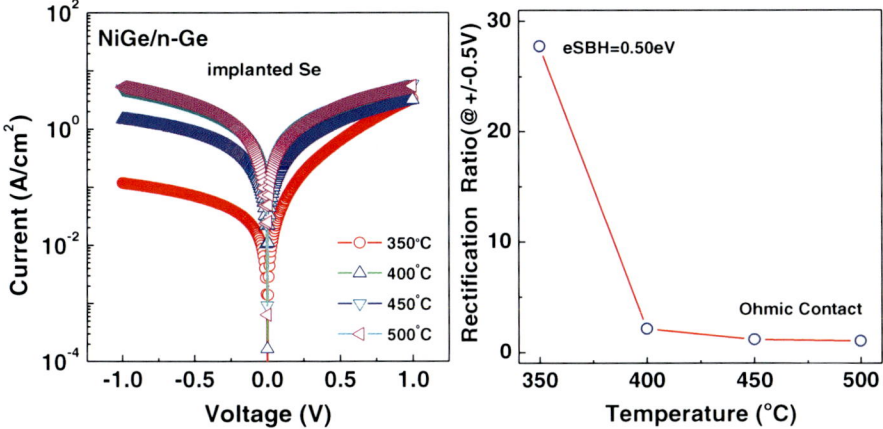

Fig. 2.16 The I–V characteristics and rectification ratio of NiGe/n-Ge (100) Schottky diodes with Se implantation as a function of annealing temperature

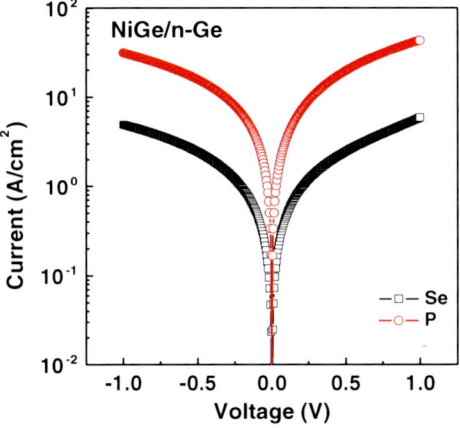

Fig. 2.17 The ohmic characteristics of NiGe/n-Ge (100) Schottky diodes with P and Se implantation

shown in Fig. 2.17, and this is because that the Se atoms may have a bad influence on the quality of NiGe film, which leads to the film resistance increase.

The SBH modulation is resulted by snowplow effect, which makes the implantation dopants pile up at the contact interface [24]. On the one hand, the segregated Se can passivate the dangling bond of Ge surface, and thus reduce the interface state density. On the other hand, shallow donor-like trap level at 0.14 eV below the conduction band of Ge is introduced by Se [25], and after these atoms are positively charged, dipoles across the interface are formed. These dipoles lead to Ge energy

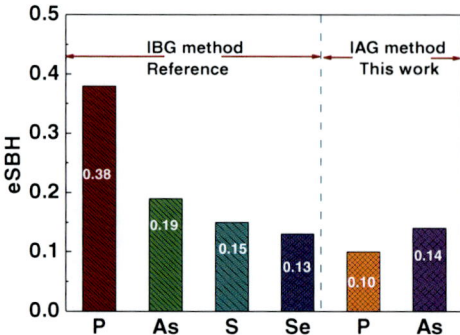

Fig. 2.18 The electron SBHs achieved by IAG and IBG techniques with different species

band bending downward, which reduces the electron barrier width for tunneling, and thus the effective electron SBH is decreased.

2.3.3   Comparison of the IAG and IBG Techniques

Figure 2.18 gives the obtained electron SBH of NiGe/Ge contact achieved by different methods. For IBG technique, the electron SBHs of 0.38 and 0.19 eV are realized by the pregermanide implantation of P and As [18], respectively. At the same time, the electron SBHs of 0.15 and 0.13 eV are obtained for the S and Se implantation [26, 27]. While for the IAG technique, lower electron SBHs of 0.10 and 0.14 eV are achieved by P and As implantation [28, 29]. Based on these results, the IAG technique shows superior advantages over IBG technique, and the reasons can be summarized as follows. First, for the IBG technique, the implanted dopant may diffuse toward the substrate during the germanidation process, which results in an inferior segregation effect compared with IAG technique. Second, the implantation damage will exacerbate the implanted dopant diffusion, and the dopant concentration at the NiGe/Ge interface is further reduced. Finally, the implanted dopant in Ge is unfavorable for NiGe formation, especially for large radius dopant. The poor quality of NiGe film will result in the increase of film resistance, further influencing device performance. While for the IAG technique, it can avoid the implantation damage and shows more process flexibility, thus it is very promising for its application in Ge-based device.

2.4 Summary

In this chapter, the electron SBH modulation of metal/Ge contact is systematically studied. The dopant segregation (including IAG and IBG methods) is mainly focused to modulate the eSBH of NiGe/Ge contact, and the process parameters of implantation and annealing are optimized.

1. The IAG technique is first applied to modulate the electron SBH of NiGe/Ge diode with P+ and As+ implantation. The current characteristics of NiGe/p-Ge diode change from ohmic to well rectifying with I_{on}/I_{off} ratio over 10^5 and 10^4, and low electron barrier height of 0.10 eV and 0.14 are achieved for P+ and As+ implantation. Besides, the current characteristics of NiGe/n-Ge diode show well ohmic characteristics, which is beneficial to promoting the performance of Ge-based SB MOSFETs.
2. The process parameters of IAG technique are optimized. For P+ implantation, the optimum implantation energy and dose are 50 keV and 1×10^{15} cm^{-2}, and the process window is 350–500 °C. While for As+ implantation, the optimum implantation energy and dose are 60 keV and 1×10^{15} cm^{-2}, and the process window is 450–500 °C. These results provide guideline for Ge-based SB MOSFET fabrication.
3. For the IBG technique, the electron SBH is effectively modulated and ohmic characteristics are realized with P and Se implantation. However, the annealing temperature should be controlled below 400 °C for P+ implantation, and the implanted Se may result in poor quality of NiGe with large film resistance.

References

1. Yamazaki S (1990) Insulated gate field effect transistor and its manufacturing method. ed: Google Patents
2. Lepselter MP, Sze SM (1968) SB-IGFET: an insulated-gate field-effect transistor using Schottky barrier contacts for source and drain. Proc IEEE 56:1400–1402
3. Ikeda K, Kamimuta Y, Moriyama Y, Ono M, Usuda K, Oda M et al (2013) Enhancement of hole mobility and cut-off characteristics of strained Ge nanowire pMOSFETs by using plasma oxidized GeOx inter-layer for gate stack. In: VLSI technology (VLSIT), 2013 symposium on 2013, pp T30–T31
4. Ikeda K, Ono M, Kosemura D, Usuda K, Oda M, Kamimuta Y et al (2012) High-mobility and low-parasitic resistance characteristics in strained Ge nanowire pMOSFETs with metal source/drain structure formed by doping-free processes. In: VLSI technology (VLSIT), 2012 symposium on, pp 165–166
5. Liu B, Gong X, Han GQ, Lim PSY, Tong Y, Zhou Q et al (2012) High-performance germanium omega-gate mugfet with Schottky-barrier nickel germanide source/drain and low-temperature disilane-passivated gate stack. IEEE Electron Device Lett 33:1336–1338
6. Xiong SY, King TJ, Bokor J (2005) A comparison study of symmetric ultrathin-body double-gate devices with metal source/drain and doped source/drain. IEEE Trans Electron Devices 52:1859–1867

7. Connelly D, Faulkner C, Grupp DE (2003) Optimizing Schottky S/D offset for 25-nm dual-gate CMOS performance. IEEE Electron Device Lett 24:411–413
8. Tersoff J (1984) Schottky barrier heights and the continuum of gap states. Phys Rev Lett 52:465–468
9. Zhou Y, Han W, Yong W, Xiu F, Zou J, Kawakami RK et al (2010) Investigating the origin of Fermi level pinning in Ge Schottky junctions using epitaxially grown ultrathin MgO films. Appl Phys Lett 96:102103–102103-3
10. Heine V (1965) Theory of surface states. Phys Rev 138:A1689
11. Nishimura T, Kita K, Toriumi A (2007) Evidence for strong Fermi-level pinning due to metal-induced gap states at metal/germanium interface. Appl Phys Lett 91:123123
12. Dimoulas A, Toriumi A, Mohney SE (2009) Source and drain contacts for germanium and III–V FETs for digital logic. MRS Bull 34:522–529
13. Dimoulas A, Tsipas P, Sotiropoulos A, Evangelou E (2006) Fermi-level pinning and charge neutrality level in germanium. Appl Phys Lett 89:252110–252110-3
14. Thornton R (1981) Schottky-barrier elevation by ion implantation and implant segregation. Electron Lett 17:485–486
15. Yamauchi T, Nishi Y, Tsuchiya Y, Kinoshita A, Koga J, Kato K (2007) Novel doping technology for a 1nm NiSi/Si junction with dipoles comforting Schottky (DCS) barrier. In: 2007 IEEE international electron devices meeting, vols 1 and 2, pp 963–966
16. Zhen Z, Qiu Z, Ran L, Ostling M, Zhang S-L (2007) Schottky-barrier height tuning by means of ion implantation into preformed silicide films followed by drive-in anneal. IEEE Electron Device Lett 28:565–568
17. Hoong-Shing W, Lap C, Samudra G, Yee-Chia Y (2007) Effective Schottky barrier height reduction using sulfur or selenium at the NiSi/n-Si (100) interface for low resistance contacts. IEEE Electron Device Lett 28:1102–1104
18. Mueller M, Zhao Q, Urban C, Sandow C, Buca D, Lenk S et al (2008) Schottky-barrier height tuning of NiGe/n-Ge contacts using As and P segregation. Mater Sci Eng B 154:168–171
19. Guo Y, An X, Huang R, Fan CH, Zhang X (2010) Tuning of the Schottky barrier height in NiGe/n-Ge using ion-implantation after germanidation technique. Appl Phys Lett 96. 5 Apr 2010
20. Guo Y, An X, Wang R, Zhang X, Huang R (2011) Investigation on morphology and thermal stability of NiGe utilizing ammonium fluoride pretreatment for germanium-based technology. IEEE Electron Device Lett 32(4):554–556
21. Sze SM, Ng KK (2006) Physics of semiconductor devices. Wiley, New Jersey
22. Simoen E, Vanhellemont J (2009) On the diffusion and activation of ion-implanted n-type dopants in germanium. J Appl Phys 106:103516
23. Zhang Z, Qiu ZJ, Liu R, Östling M, Zhang S-L (2007) Schottky barrier height tuning by means of ion implantation into pre-formed silicide films followed by drive-in anneal. IEEE Electron Device Lett 28(7):565–568
24. Wittmer M, Seidel T (2008) The redistribution of implanted dopants after metal-silicide formation. J Appl Phys 49:5827–5834
25. Sze S, Irvin J (1968) Resistivity, mobility and impurity levels in GaAs, Ge, and Si at 300 K. Solid-State Electron 11:599–602
26. Tong Y, Liu B, Lim PSY, Yeo Y-C (2012) Selenium segregation for effective Schottky barrier height reduction in NiGe/n-Ge contacts. IEEE Electron Device Lett 33:773–775
27. Ikeda K, Yamashita Y, Sugiyama N, Taoka N, Takagi SI (2006) Modulation of NiGe/Ge Schottky barrier height by sulfur segregation during Ni germanidation. Appl Phys Lett 88:152115–152115-3
28. Li Zhiqiang, An Xia, Li Min, Yun Quanxin, Lin Meng, Li Ming, Zhang Xing, Huang Ru (2012) Low electron Schottky barrier height of NiGe/Ge achieved by ion-implantation after germanidation technique. IEEE Electron Device Lett 33(12):1687–1689
29. Li Z, An X, Li M, Yun Q, Lin M, Li M, Zhang X, Huang R (2012) Study on Schottky barrier modulation of NiGe/Ge by ion-implantation after germanidation technique. In: The 11th ICSICT, Xi'an

Chapter 3
Metal Germanide Technology

Ge is considered as a potential channel material for high-performance CMOS device at future technology node for its high and more symmetric carrier mobility. In order to reduce parasitic resistance in source/drain, a self-aligned germanide process, like self-aligned silicide, is inevitable in high-performance Ge MOSFETs. To date, germanides of metals such as Ni, Pt, Ti, and Co have been investigated and the material properties of these germanides are given in Table 3.1. Compared with other germanides, NiGe is highly attractive due to its low formation temperature (about 250 °C), low resistivity (15 $\mu\Omega$ cm), and small Ge consumption [1, 2].

However, some problems need to be addressed for NiGe application in Ge MOS device. In addition to the severe Fermi level pinning effect, the poor thermal stability of NiGe is another limiting factor for device performance improvement. NiGe is formed by the reaction of Ni and Ge with rapid thermal annealing [3]. At temperature below 350 °C, the Ni-rich phases Ni_5Ge_3/Ni_2Ge are formed. Starting with 350 °C, the NiGe phase dominates. With annealing temperature over 500 °C, the quality of NiGe begins to degenerate due to agglomeration, which results in the increase of film resistance. It is well known that the thermal degradation of NiSi at high temperature includes two mechanisms: agglomeration and phase transformation [4]. In contrast, the NiGe shows a different degradation characteristic with NiSi [5]. The nickel germanide does not transform to Ge-rich phase after annealing since no Ge-rich phase (>50 at. % Ge) exists in the equilibrium phase diagram. Morphological degradation resulted by agglomeration is the only mechanism of thermal degradation in NiGe.

To date, two methods have been proposed to improve the thermal stability of NiGe film. One is incorporating other metals to form nickel alloy, such as Pt, Ti, Ta, and Yb [6–10]. These incorporated metals result in a decrease of the Gibbs energy and thus increase in the stabilization of NiGe. The other method is adopting TiN as a capping layer to reduce grain boundary, this method is also beneficial for thermal stability improvement [11]. However, these two methods can only enhance the

© Springer-Verlag Berlin Heidelberg 2016
Z. Li, *The Source/Drain Engineering of Nanoscale Germanium-based MOS Devices*, Springer Theses, DOI 10.1007/978-3-662-49683-1_3

Table 3.1 material properties of germanides

Germanide	Formation temperature (°C)	Resistivity ($\mu\Omega$ cm)	Germanide/metal ratio
$TiGe_2$	800	20	–
$CoGe_2$	425	35	–
NiGe	270	15	2.55
$PtGe_2$	475	28.2	3.67

thermal stability of NiGe to 550 °C. Therefore, the P and Sb co-implantation technique is proposed to further improve the thermal stability of NiGe film in this thesis.

3.1 Thermal Stability Improvement

The morphology of NiGe film with P and Sb co-implantation is characterized by scanning electron microscope (SEM) and atomic force microscope (AFM) measurements, and these results demonstrate that the thermal stability of NiGe film is greatly improved by this technique.

3.1.1 Experiment Design

Fabrication of devices were carried out on p-type and n-type Ge (100) substrates with a resistivity of 1–10 Ω cm. First, 300 nm SiO_2 film was deposited on the Ge substrates by plasma enhanced chemical vapor deposition and patterned by the conventional lithography to define the diode areas. After being pretreated with ammonium fluoride pretreatment (AFP) method, 20 nm Ni was sputtered and followed by a rapid thermal annealing at 400 °C in nitrogen ambient to form the NiGe film. Then, three ion implantation conditions with P (50 keV, 1×10^{15} cm^{-2}), Sb (65 keV, 1×10^{15} cm^{-2}), and P/Sb (50 keV, 1×10^{15} cm^{-2}/65 keV, 1×10^{15} cm^{-2}) were performed to these samples, respectively, and followed by post-germanidation annealing ranging from 400–600 °C for 60 s in nitrogen ambient. Finally, the substrate contact was prepared directly through thermal evaporation of Al onto the Ge backside. The control sample, without ion implantation, was also fabricated for comparison.

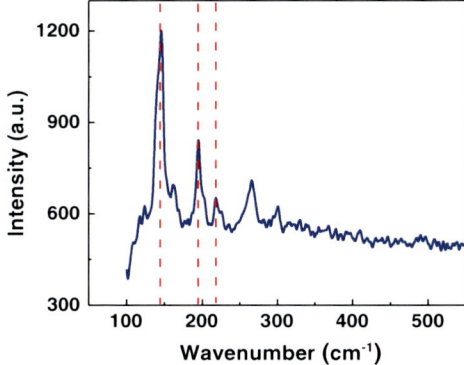

Fig. 3.1 Micro-Raman spectrum of NiGe films formed at 400 °C

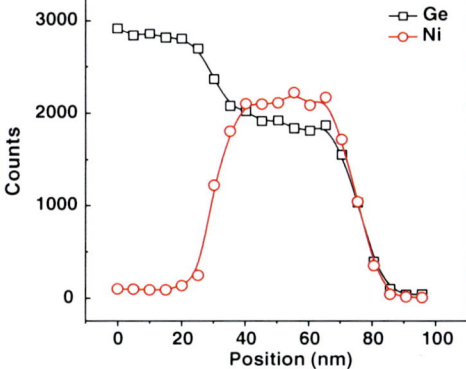

Fig. 3.2 The energy dispersive X-ray spectroscopy analysis of NiGe film formed at 400 °C

3.1.2   Property Analysis

Figure 3.1 shows the micro-Raman spectrum of nickel germanide films formed at 400 °C without ion implantation. The sample shows the Raman peaks at 140, 194, and 217 cm^{-1}, which are the typical Raman spectrum of NiGe [12]. Meanwhile, the energy dispersive X-ray spectroscopy (EDX) of NiGe film in Fig. 3.2 shows that the ratio of Ni and Ge atoms is about 1, thus further verifying the NiGe phase. For the sample annealed at 500 °C for 60 s, a Raman peaks at 300 cm^{-1} is observed in Fig. 3.3, which belongs to the phase of Ge substrate. This is because an "island" structure is formed due to NiGe aggregation at 500 °C, which is consistent with the published results [13].

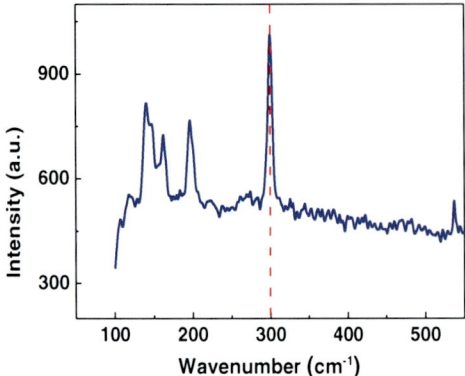

Fig. 3.3 Micro-Raman spectrum of NiGe films formed at 500 °C

3.1.3 Morphology Improvement of NiGe Film

Figure 3.4 shows the comparison of scanning electron microscopy (SEM) images of P and P+Sb implanted samples. The NiGe film keeps continuous and smooth for the pure NiGe sample formed at 400 °C. In the P implanted sample, NiGe film begins to degrade after postgermanidation annealing at 500 °C. While for the P+Sb implanted sample, the surface of NiGe film still keeps smooth without agglomeration after postgermanidation annealing at 500 °C, indicating the well improved morphology of NiGe film.

The surface morphology of the implanted samples after post-germanidation anneal at 500 °C for 60 s was studied by atomic force microscopy (AFM) measurement [14]. The P+Sb implanted sample exhibits small root mean square (RMS) roughness of 1.70 nm, which is smaller than that of the P implanted sample (3.15 nm), and this is consistent with the SEM results. Besides, the only Sb implanted sample also shows a similar RMS of 1.69 nm, which demonstrates that the beneficial impacts of Sb are responsible for the improvement.

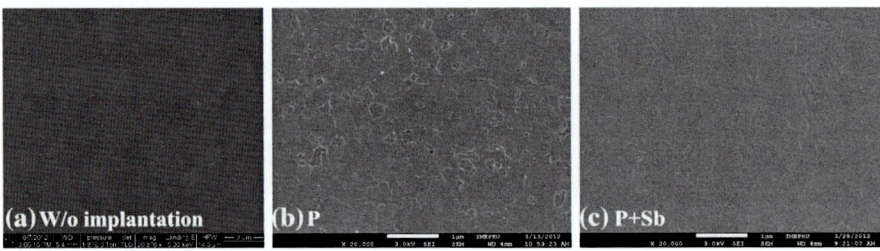

Fig. 3.4 The SEM images of NiGe film with different implanted conditions **a** w/o implantation **b** P implantation. **c** P+Sb implantation

3.1.4 Thermal Stability Improvement of NiGe Film

Figure 3.5 shows the comparison of scanning electron microscopy (SEM) images of P and P+Sb implanted samples. In the P implanted sample, obvious aggregation phenomenon of NiGe film occurred at 500 °C, while for the P+Sb and Sb implanted samples, the NiGe film still keeps flat and continuous when the annealing temperature increases to 600 °C. Therefore, the thermal stability of NiGe film is enhanced to at least 600 °C with the P and Sb co-implantation technique.

The combined effects of surface passivation and strain relaxation in the presence of Sb are probably responsible for the thermal stability improvement of NiGe film. The Sb atoms arrived at the NiGe/Ge interface can form Ge(001)−(2 × 1)-Sb reconstruction, and thus the surface dangling bonds on Ge substrate are terminated, which leads to a passivated Ge surface [15]. In addition, the maximum lattice mismatch between NiGe and Ge is 14.25 % [16], which is likely to result in strain. The strain can be relieved by Sb elements existing at the interface, which is beneficial to form an abrupt contact interface, and this is confirmed by the cross-sectional transmission electron microscopy images of P and P+Sb implanted samples with post-germanidation annealing at 500 °C. A defect-rich region near the interface of the P implanted sample is observed in Fig. 3.6, whereas the P+Sb implanted sample

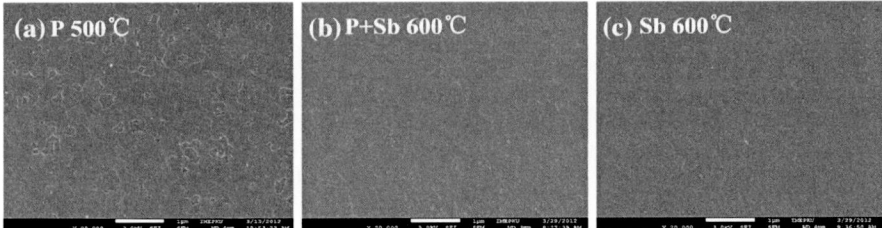

Fig. 3.5 The SEM images of NiGe with implanting P and P+Sb at various post-germanidation annealing temperatures

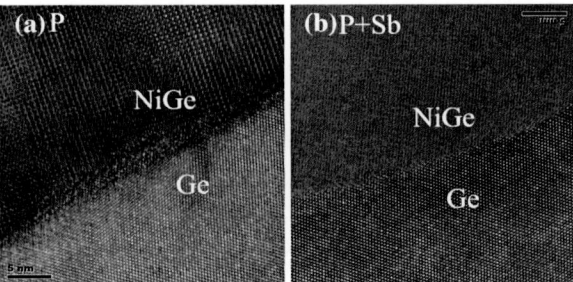

Fig. 3.6 The cross-sectional TEM images of P and P+Sb implanted samples with the same post-germanidation annealing at 500 °C

exhibits a smooth and clear NiGe/Ge contact interface, without disorder or inter-mixing existing at the interface. Besides, the growth of NiGe grain is restrained by the Sb element as shown in Fig. 3.6, which also contributes to enhance the thermal stability of NiGe film.

3.2 Electrical Performance Improvement of NiGe/Ge Contact

The impact of P and Sb co-implantation technique on the electrical characteristic of NiGe/Ge contact is investigated in this part.

3.2.1 NiGe/Ge Schottky Diodes

Figure 3.7 shows the I-V characteristics of the NiGe/Ge Schottky diodes with P and Sb co-implantation technique. The P+Sb implanted sample of NiGe/n-Ge contact shows better ohmic characteristic than that of the P implanted sample. Besides, the reverse current of P+Sb implanted sample is further reduced by over one order of magnitude compared to the P implanted sample. Therefore, the electrical character-istic of the NiGe/Ge contact has also been greatly improved by the P and Sb co-implantation technique.

The improved electrical characteristics of the NiGe/Ge contact can be ascribed to higher dopant concentration near the interface. Due to simultaneous segregation of P and Sb near the contact interface, the local tensile strain resulted by P element is counteracted by the compressive strain introduced by Sb element, and thus the

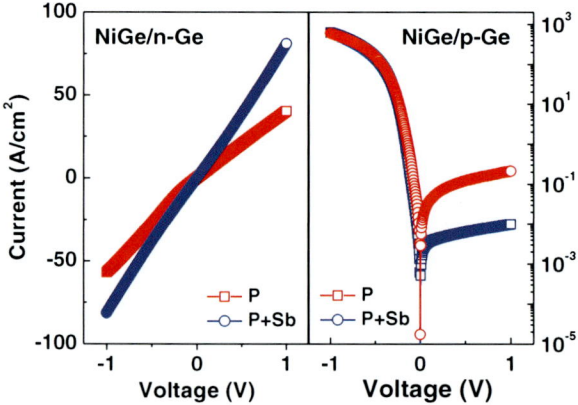

Fig. 3.7 The I–V curves of NiGe/Ge Schottky diodes with P and P+Sb implantation

effective solid solubility of implanted n-type dopants is increased according to the first Hume–Rothery's rule [17]. The higher dopant concentration near the interface is expected to increase the effective hole Schottky barrier, and thereby a lower reverse current is obtained. In addition, the Ge surface can be effectively passivated by Sb, and thus an improved contact interface is obtained, which is beneficial for reducing leakage current.

3.2.2   NiGe/n-Ge Contact Resistivity

Since the electrical characteristic of the NiGe/Ge contact is improved by P and Sb co-implantation technique, the contact resistivity of NiGe/Ge contact is also extracted by circular transmission line model (CTLM) in this thesis [18]. The process flow of the measure structure is given in Fig. 3.8. The p-type and n-type Ge(100) substrates with resistivity of 1–10 Ω cm were used to measure structure fabrication. After cleaning the wafers, 20 nm Ni was sputtered and followed by rapid thermal annealing at 400 °C in nitrogen ambient to form NiGe film. Then, the samples were divided into two groups for ion implantation with P (50 keV, 1×10^{15} cm^{-2}) and P/Sb (50 keV, 1×10^{15} cm^{-2}/65 keV, 1×10^{15} cm^{-2}), respectively. Post-germanidation annealing was performed for all the samples at 500 °C for 60 s in nitrogen ambient. Finally, 100 nm Al layer was sputtered on NiGe as electrode.

Figure 3.9 gives the CTLM resistance as a function of gap spacing with the P and Sb co-implantation technique. The CTLM resistance increases linearly with gap spacing d, which means that the measured results accord with the CTLM well. The contact resistivity of P and P+Sb implanted samples is compared in Fig. 3.10. The typical value of NiGe/n-Ge contact formed at 400 °C is about 8×10^{-5} Ω cm^2 [19],

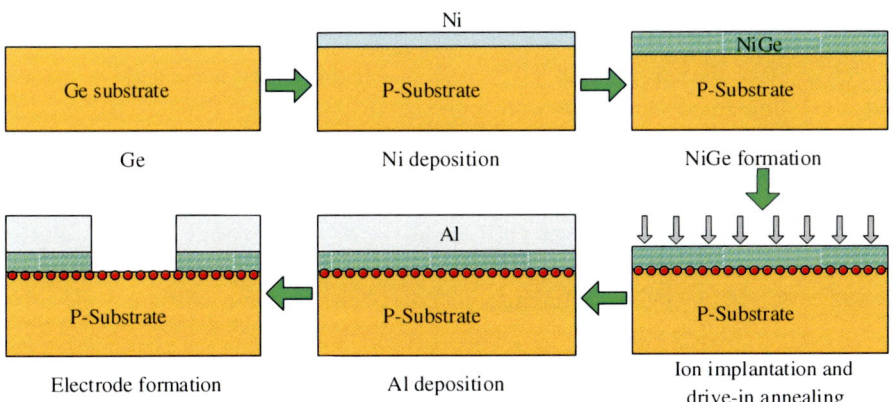

Fig. 3.8 The process flow of the measure structure with P+Sb co-implantation technique

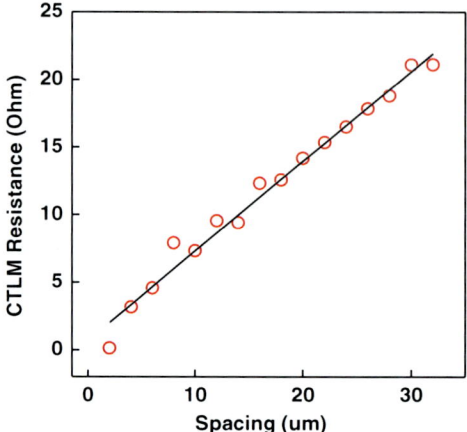

Fig. 3.9 Linear fitting the CTLM resistance with gap spacing for the P+Sb implanted sample

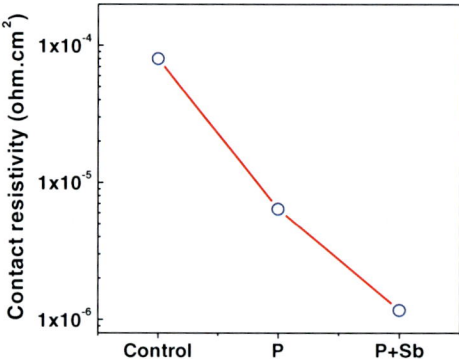

Fig. 3.10 The contact resistivity of NiGe/n-Ge contact with P and P+Sb implantation

which is due to the severe Fermi level pinning effect between NiGe and Ge contact. For the P implanted sample, the contact resistivity is reduced to 6.4×10^{-6} Ω cm^2, owing to the obvious electron SBH reduction resulted by P segregation at the NiGe/Ge contact interface. While for the P+Sb implanted sample, the contact resistivity is further reduced to 1.2×10^{-6} Ω cm^2, this is due to the more obvious SBH modulation by the higher dopant concentration at the NiGe/Ge contact interface. Therefore, the contact resistivity of NiGe/n-Ge contact can be effectively reduced by the P and Sb co-implantation technique, which is beneficial for enhancing the drive current of Ge nMOS device.

3.3   Ge-based Schottky Barrier Source/Drain nMOSFET

Since both the thermal stabilities of NiGe and the electrical characteristics NiGe/Ge contact are significantly enhanced by P and Sb co-implantation, it is expected that the P and Sb co-implantation technique will have a positive influence on Ge-based SB nMOSFET, and the following part will investigate its influence on the performance of Ge-based SB nMOSFET.

3.3.1   Device Fabrication

Figure 3.11 gives the main process flow of Ge-based SB nMOSFETs. This fabrication step employed p-type (100) Ge wafers with resistivity of $1-10 \, \Omega$ cm. After HCl rinsing and surface oxide removing, the Ge wafers were loaded into atomic layer deposition (ALD) equipment, and then nitrogen plasma treatment was carried out to passivate the Ge wafer surface. After that, 5 nm Al_2O_3 gate dielectric was deposited by ALD and 100 nm TiN was sputtered, followed by inductively coupled plasma (ICP) etch to form a metal gate. Then, 50 nm of SiO_2 was deposited and etched to form a sidewall. After precleaning of Ge wafers, 20 nm Ni was sputtered and followed by rapid thermal annealing at 400 °C in nitrogen ambient to form a NiGe film. The unreacted Ni was removed by HCl/H_2O solution. The P dopants (50 keV, 1×10^{15} cm^{-2}) and Sb dopants (65 keV, 1×10^{15} cm^{-2}) were then implanted in the metal S/D regions. The drive-in annealing was carried out at 500 °C for 60 s in N_2 for dopants segregated at the NiGe/p-Ge interface. After defining the S/D contact holes, Ti/Al metallization was performed to form the metal pads. The substrate contact was prepared directly through thermal evaporation of Al onto the Ge backside. Finally, the overall MOSFET structures were subjected to alloying annealing in N_2 ambient at a temperature of 375 °C for 30 min. In addition, the P+Sb implantated S/D Ge nMOSFETs were also fabricated for comparison.

3.3.2   Device Characteristic Analysis

Figure 3.12 shows the I_d–V_g transfer characteristics of the Ge nMOSFET with a NiGe Schottky junction S/D in the linear region at $V_{ds} = 0.1$ V and saturation region at $V_{ds} = 1$ V. A relatively good transfer characteristic is observed, but for exhibiting high leakage current and the reasons can be ascribed as follows. First, the small bandgap of Ge material is responsible for the leakage current of Ge-based device. Second, over-etching of the TiN layer results in damages in Ge substrate, leading to a poor interface between NiGe and Ge substrate. Finally, the HCl rinsing and surface oxide removing are not suitable for NiGe formation in gate-first process, which

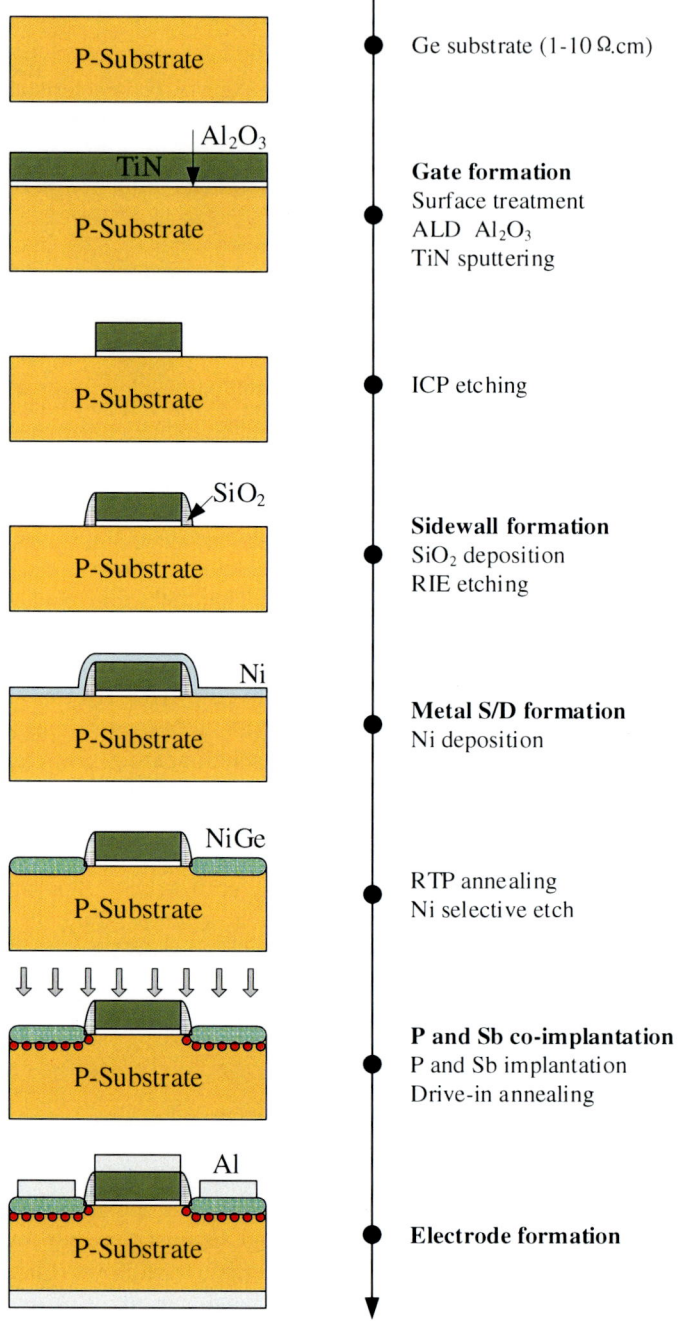

Fig. 3.11 The process flow of Ge-based SB nMOSFETs

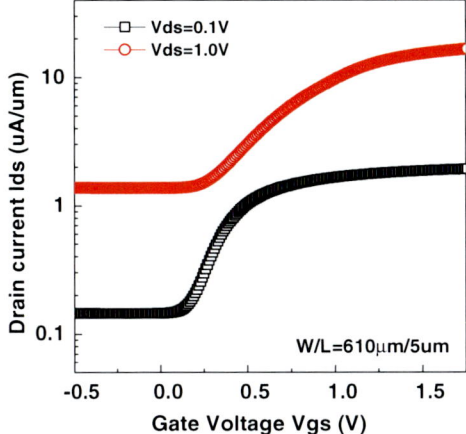

Fig. 3.12 Transfer characteristics of Ge nMOSFET with a NiGe Schottky junction S/D

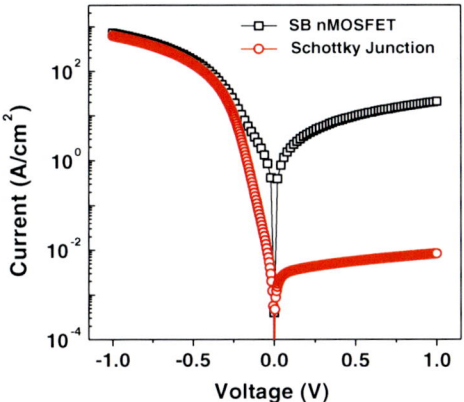

Fig. 3.13 The I–V characteristics of NiGe/p-Ge contact in Schottky diodes and SB nMOSFET

will deteriorate the NiGe/Ge contact interface. The I–V characteristics of NiGe/p-Ge contact with P and Sb co-implantation in previous Schottky diodes and the S/D junction of Ge SB nMOSFET are compared in Fig. 3.13. As shown, the reverse current of the S/D junction in Ge SB nMOSFET is 3 magnitudes larger than that of previous Schottky diodes, which verifies that the gate-first process is unfavorable for the junction characteristic of NiGe/Ge contact.

The I_d–V_{ds} output characteristics of Ge-based SB nMOSFETs and implantated S/D nMOSFETs are given in Figs. 3.14 and 3.15. The saturation current of Ge SB nMOSFETs (17.6 µA/µm, Vgs = 1.5 V) corresponds to the implantated S/D nMOS-FETs (19.3 µA/µm, Vgs = 1.5 V). But implantated S/D nMOSFETs shows a slow-saturation behavior, indicating a large series resistance in S/D. In conclusion, the P

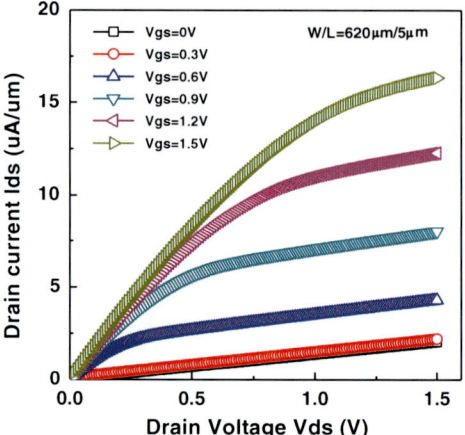

Fig. 3.14 The Id–Vds output characteristics of Ge SB nMOSFETs

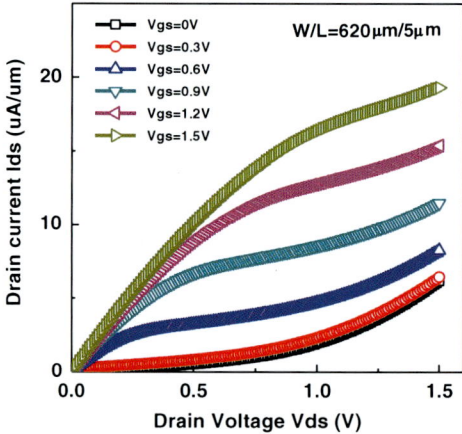

Fig. 3.15 The Id–Vds output characteristics of Ge implantated S/D nMOSFETs

and Sb co-implantation is beneficial to the performance improvement of Ge-based device, but the gate-last process may be more suitable for its application in Ge-based SB nMOSFETs.

3.4 Summary

NiGe is considered as the most promising germanide material candidate because of its low formation temperature, low resistivity, and the feasibility for self-aligned fabrication. However, the poor thermal stability of NiGe is an important limiting factor for its application in Ge MOSFET. Therefore, the P and Sb co-implantation technique is proposed in this thesis to improve the thermal stability of NiGe film, the morphology and electrical performance improvement of NiGe/Ge Contact by P and Sb co-implantation are experimentally demonstrated.

(1) The morphology of NiGe is greatly improved by P and Sb co-implantation technique with a reduced RMS roughness of 1.70 nm. Meanwhile, the thermal stability is enhanced to at least 600 °C, providing a wide temperature window for Ge-based MOS device fabrication.
(2) The quality improvement of NiGe film is ascribed to the combined effects of surface passivation and strain relaxation in the presence of Sb.
(3) The electrical characteristics of the NiGe/Ge contact is improved by P and Sb co-implantation technique, behaving as a low leakage current of NiGe/p-Ge Schottky diodes and a good ohmic characteristic of NiGe/n-Ge contact with a small resistivity of $1.2 \times 10^{-6} \, \Omega \, cm^2$.
(4) The P and Sb co-implantation technique is successfully integrated into Ge-based SB nMOSFETs and the device offers well-behaved output characteristics.

References

1. Lee K, Liew S, Chua S, Chi D, Sun H, Pan X (2004) Formation and morphology evolution of nickel germanides on Ge (100) under rapid thermal annealing. In: MRS Proceedings, p C2. 4
2. Ashburn SP, Öztürk MC, Harris G, Maher DM (1993) Phase transitions during solid-state formation of cobalt germanide by rapid thermal annealing. J Appl Phys 74:4455–4460
3. Mueller M, Zhao Q, Urban C, Sandow C, Buca D, Lenk S et al (2008) Schottky-barrier height tuning of NiGe/n-Ge contacts using As and P segregation. Mater Sci Eng B 154:168–171
4. Colgan E, Mäenpää M, Finetti M, Nicolet M (1983) Electrical characteristics of thin Ni_2Si, NiSi, and $NiSi_2$ layers grown on silicon. J Electron Mater 12:413–422
5. Nash A, Nash P (1987) The Ge-Ni (Germanium-Nickel) system. Bull Alloy Phase Diagrams 8:255–264
6. Park K, Lee B, Lee D, Ko D-H, Kwak K, Yang C-W et al (2007) A study on the thermal stabilities of the NiGe and $Ni_{1-x}Ta_xGe$ systems. J Electrochem Soc 154:H557–H560
7. Zhang Y-Y, Oh J, Li S-G, Jung S-Y, Park K-Y, Shin H-S et al (2009) Ni Germanide utilizing Ytterbium interlayer for high-performance Ge MOSFETs. Electrochem Solid-State Lett 12:H18–H20
8. Zhu S, Yu M, Lo G, Kwong D (2007) Enhanced thermal stability of nickel germanide on thin epitaxial germanium by adding an ultrathin titanium layer. Appl Phys Lett 91:051905
9. Nakatsuka O, Suzuki A, Sakai A, Ogawa M, Zaima S (2007) Impact of Pt incorporation on thermal stability of NiGe layers on Ge (001) substrates. In: 2007 International workshop on junction technology, pp 87–88
10. Mueller M, Zhao OT, Urban C, Sandow C, Breuer U, Mantl S (2008) Schottky-barrier height tuning of Ni and Pt germanide/n-Ge contacts using dopant segregation. In: 2008 9th International conference on solid-state and integrated-circuit technology, vols 1–4, pp 153–156
11. Liu J, Wen H, Lu J, Kwong D (2005) Improving gate-oxide reliability by TiN capping layer on NiSi FUSI metal gate. IEEE Electron Device Lett 26:458–460

12. Jin LJ, Pey KL, Choi WK, Fitzgerald EA, Antoniadis DA, Pitera AJ et al (2004) The interfacial reaction of Ni with (111)Ge, (100)Si$_{0.75}$Ge$_{0.25}$ and (100)Si at 400 °C. Thin Solid Films 462:151–155
13. Zhu S, Nakajima A (2005) Annealing temperature dependence on nickel–germanium solid-state reaction. Jpn J Appl Phys 44:L753
14. Li Z, An X, Li M, Yun Q, Lin M, Li M, Zhang X, Huang R (2013) Morphology and electrical performance improvement of NiGe/Ge contact by P and Sb co-implantation. IEEE Electron Device Lett 34(5):596–598
15. Rich D, Miller T, Chiang T-C (1990) Electronic and chemical properties of In and Sb adsorbed on Ge (100) studied by synchrotron photoemission. Phys Rev B 41:3004
16. Horn-von Hoegen M, LeGoues F, Copel M, Reuter M, Tromp R (1991) Defect self-annihilation in surfactant-mediated epitaxial growth. Phys Rev Lett 67:1130
17. Hume-Rothery W, Smallman RW, Haworth CW (1969) The structure of metals and alloys, Metals and Metallurgy Trust, London, UK
18. Schroder DK (2006) Semiconductor material and device characterization. Wiley
19. Oh JH, Chen Y-T, Ok I, Jeon K, Lee S-H (2010) High specific contact resistance of ohmic contacts to n-Ge source/drain and low transport characteristics of Ge nMOSFETs. In: International conference on solid state devices and materials, Japan, pp 3–20

Chapter 4
Contact Resistance of Ge Devices

As device dimension is scaling down, the source/drain parasitic resistance R_{series} makes up a growing percentage of the on-resistance ($R_{on} = V_{dd}/I_{on}$), thus it presents a larger influence on device performance [1]. However, it is very difficult to reduce R_{series} in sub-100 nm MOSFETs. Figure 4.1 shows the projected maximum R_{series} values versus the year of production in the near future that allow for the required I_{on} as set by ITRS [2]. In future years, the R_{series} is expected to be controlled below 25 % of the total R_{on} for the required I_{on}, yet no known solutions are available to meet these projections. This problem is more serious in the Ge nMOSFETs, due to the high R_{series} resulted by the severe Fermi level pinning effect and low dopant activation concentration. Therefore, R_{series} reduction presents one of the biggest challenges for the source/drain engineering of Ge nMOSFETs.

4.1 S/D Parasitic Resistance

As shown in Fig. 4.2, R_{series} can be divided into four components [3]: (i) extension-to-gate overlap resistance (R_{ov}); (ii) S/D extension resistance (R_{ext}); (iii) deep S/D resistance ($R_{S/D}$); and (iv) contact resistance at the silicide–silicon interface (R_{co}).

The summation of R_{ext} and $R_{S/D}$ is commonly referred as the spreading resistance under the sidewall spacer (R_{spr}) and the critical point for R_{spr} reduction is increasing the dopant activation concentration in the S/D regions. The contact resistance (R_{co}) is defined as the resistance between the silicide and the Si underneath the silicide and is expressed as follows:

$$R_c = \rho_c / WL_{Cont}, \quad \rho_c \propto \exp\left(\frac{2\phi_B}{\hbar}\sqrt{\frac{\varepsilon_s m^*}{N}}\right) \tag{4.1}$$

© Springer-Verlag Berlin Heidelberg 2016
Z. Li, *The Source/Drain Engineering of Nanoscale Germanium-based MOS Devices*, Springer Theses, DOI 10.1007/978-3-662-49683-1_4

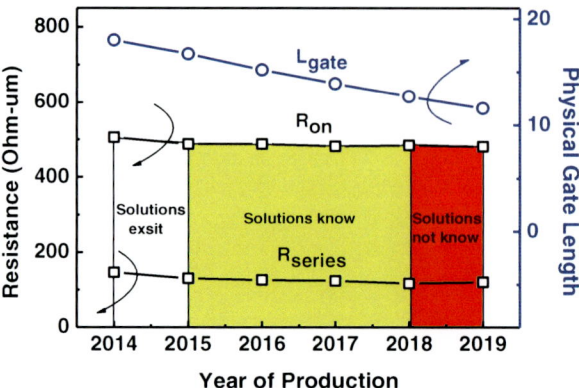

Fig. 4.1 Scaling projections of R_{on} and R_{series} for high-performance bulk nMOSFETs

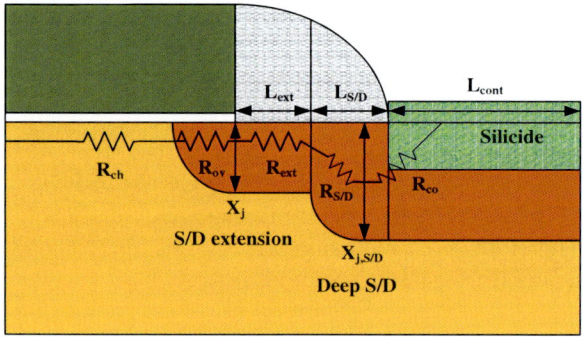

Fig. 4.2 Schematic representation of various components of S/D series resistance

where L_{Cont} is the length of the silicide, ϕ_B is the Schottky barrier height of the silicide-Si contact, m^* is the effective mass of the carriers tunneling across the contact, N is the active dopant concentration of the Si region at the interface, $\hbar$ is the Planck constant, and ε_s is the dielectric constant of Si. It can be concluded that the ρ_c depends strongly on ϕ_B and N.

Figure 4.3 gives the results of an advanced series resistance model for sub-100 nm bulk-Si technologies [3–5]. As it is shown, R_{co} is the dominant components and it accounts for >65 % of the total series resistance, thus reducing R_{co} will play a critical role in the device performance improvement. Figure 4.4 shows the projected R_{co} values in the near future technology node set by ITRS. With device dimensions scaling down to 20 nm, a low contact resistivity below 10^{-8} Ω cm² is needed to meet the requirement of ITRS. For the Ge-based device, the lowest contact resistivity of metal/p-Ge contact is 2.7×10^{-8} Ω cm² [6], while that of metal/n-Ge contact is still limited of the order of 10^{-6} Ω cm² due to low n-type dopant concentration and large

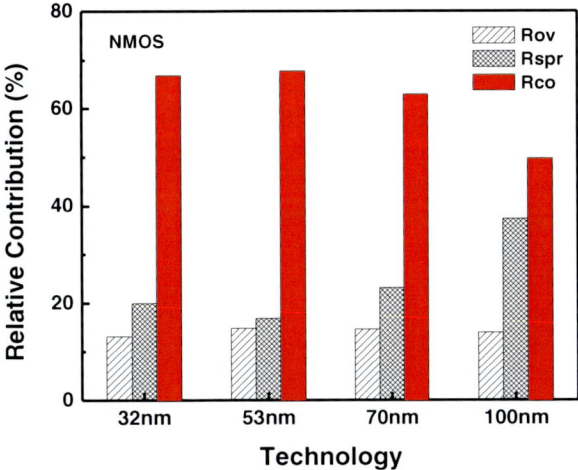

Fig. 4.3 The relative contribution of the various components of the series resistance for different technology nodes

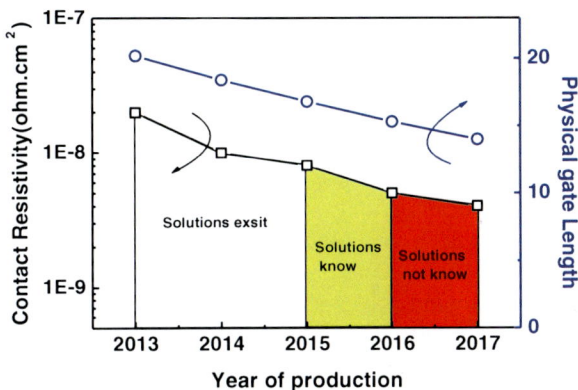

Fig. 4.4 Scaling projections of contact resistivity for high-performance bulk nMOSFETs

electron SBH [7]. The electron SBH modulation techniques are systematically studied in Chap. 2 and this chapter will focus on the n-type dopant activation enhancement to reduce the S/D contact resistance of Ge-based nMOSFETs.

4.2   Multiple Implantation and Multiple Annealing Technique

The dopant activation concentration has an important influence on the source/drain contact resistance of Ge-based MOS device. To date, a high p-type dopant electrical concentration in Ge over 1×10^{20} cm^{-3} is achieved but the electrical concentration of n-type dopant is still limited around 2×10^{19} cm^{-3} [8, 9], which results in high contact resistance in the S/D of Ge nMOSFETs. The mechanism responsible for the low electrical concentration of n-type dopant in Ge is the damages introduced by ion implantation. These damages not only accelerate dopant diffusion but also act as acceptor state in the bandgap and compensate the donor [8]. It has been reported that the implantation-induced damage can be reduced with multiple implantation and multiple annealing (MIMA) technique, and thus a high electrical concentration of n-type dopant is obtained [10]. Therefore, the MIMA technique is adopted in this chapter to reduce the contact resistivity of metal/n-Ge contact, and its impact on the performance of Ge n+/p junction is also investigated.

4.2.1   Experiment Design

The process flow of device fabrication is given in Fig. 4.5. Fabrication of devices was carried out on p-type Ge (100) substrate. The p-type Ge (100) substrates were used for the device fabrication. These wafers were divided into two groups for the circular transfer length method (CTLM) structure and Ge n+/p diodes fabrication, respectively. First, 300 nm SiO$_2$ film was deposited on the Ge substrates and patterned by conventional lithography to define the diode areas for Ge n+/p diodes fabrication. Then, 20 nm SiO$_2$ was deposited onto all the samples as a cap layer to reduce the implantation damage and prevent dopant loss during dopant activation. After that, the P$^+$ was implanted at energy of 50 keV with a dose of 5×10^{14} cm^{-2}, followed by rapid thermal annealing at 500 °C for 30 s in nitrogen ambient. This process was implemented multiple times. Finally, contact openings were defined by conventional lithography followed by Ti/Al (5 nm/100 nm) deposition and liftoff to form contacts.

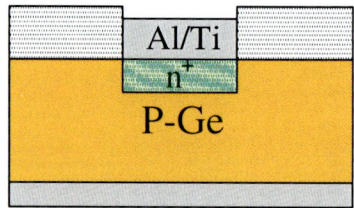 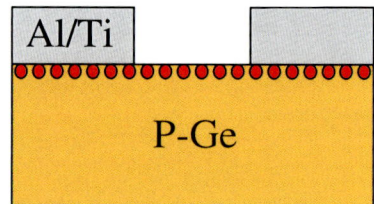

Fig. 4.5 The structures of Ge n+/p junction and CTLM with MIMA technique

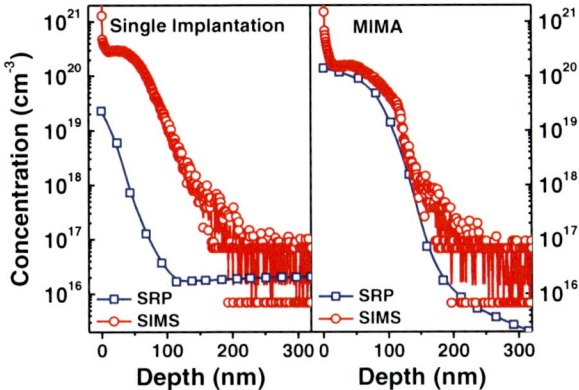

Fig. 4.6 The SIMS and SRP profiles of MIMA and single-implantation samples

4.2.2 N-type Dopant Activation

The chemical and electrical concentration profiles are measured by secondary ion mass spectrometry (SIMS) and spreading resistance profiling (SRP) analysis. As shown in Fig. 4.6, the chemical concentrations are both over 1×10^{20} cm^{-3} with the total implantation dose of 2×10^{15} cm^{-2} for the MIMA sample and single-implantation sample. The chemical concentration of MIMA is slightly smaller than that of the single-implantation sample due to longer annealing time. However, the electrical concentration of single implanted sample is 1×10^{19} cm^{-3}, which is consistent with the reported value activated by RTP annealing [8]. On the contrary, a high electrical concentration over 1×10^{20} cm^{-3} is achieved for the MIMA sample, and the much higher electrical concentration of the MIMA sample demonstrates that the MIMA technique can effectively improve the n-type dopant activation concentration in Ge.

The small implantation damages achieved by the MIMA technique are responsible for the improved donor activation. By dividing the single-implantation step into several smaller implantation cycles, the implantation damage in every cycle can be repaired by every annealing cycle, thus the total implantation damage is reduced without reducing total implantation dose, which is confirmed by Raman analysis. As shown in Fig. 4.7, with the same total implantation dose of 2×10^{15} cm^{-2}, the Raman intensity peak of the MIMA sample is higher than that of the single-implantation sample, which demonstrates that the implantation damages are effectively reduced by the MIMA technique [11].

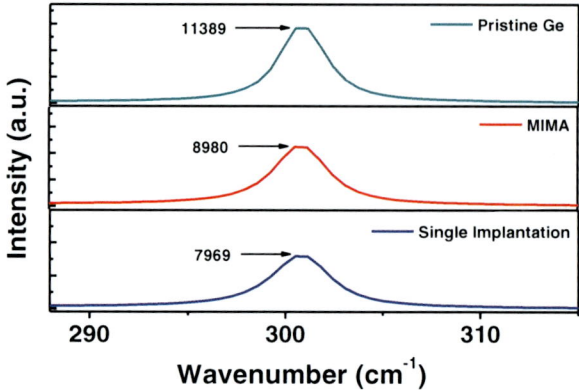

Fig. 4.7 Raman spectroscopy obtained from the samples fabricated with MIMA and single-implantation technique

4.2.3   Contact Resistivity of Metal/n-Ge Contact

The contact resistivity of metal/n-Ge contact with MIMA technique is also extracted with CTLM structure [12]. As shown in Fig. 4.8, the CTLM structure has an inner contact (100 μm radius) and an outer contact with variable gap spacing between them (from 2 to 32 μm).

The total resistance R_T between the two contacts can be expressed as follows:

$$R_{\mathrm{T}} = \frac{R_{\mathrm{sh}}}{2\pi}\left[\ln\left(\frac{R+d}{R}\right) + L_{\mathrm{T}}\left(\frac{1}{R} + \frac{1}{R+d}\right)\right] \tag{4.2}$$

where R_{sh} is the sheet resistance of the underlying doped layer (i.e., n+ Ge), L_T is the transfer length. If $L \gg d$, the expression of R_T can be simplified as follows:

$$R_{\mathrm{T}} = \frac{R_{\mathrm{sh}}}{2\pi R}\left(d + 2L_{\mathrm{T}}\right)C \tag{4.3}$$

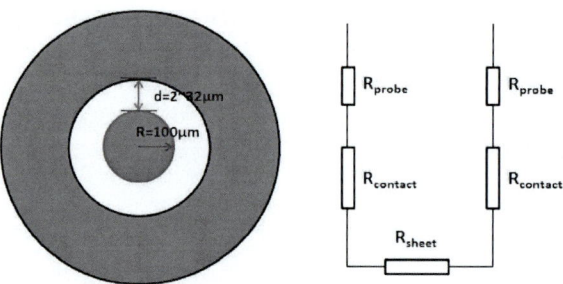

Fig. 4.8 The schematic of CTLM structure and its equivalent circuit

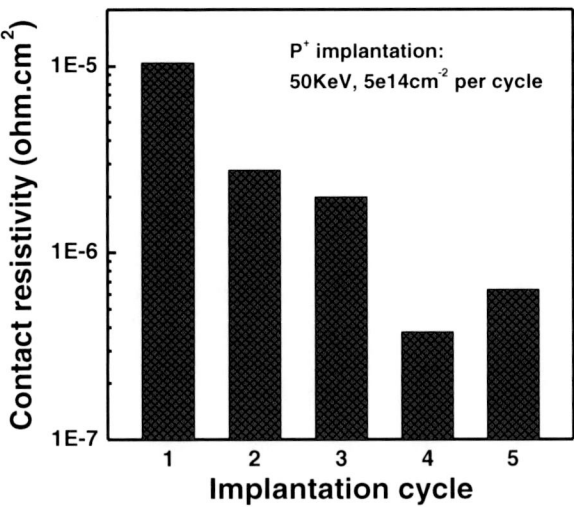

Fig. 4.9   The ρ_c of metal contact to n-Ge with increasing implantation cycle for the MIMA samples

where C is a correction factor due to the circular geometry, $C = \frac{R}{d}\ln\left(1 + \frac{d}{R}\right)$. Using this correction factor, R_T will be a linear function of gap spacing d. By plotting R_T versus gap spacing d, the R_{sh} and L_T are extracted from the slope and the y-intercept, respectively. Finally, the specific contact resistivity can be estimated by $\rho_c = R_{sh}L_T^2$.

Besides, the 4-point probe setup is used to negate the effect of the probe resistance on the measurement.

The extracted ρ_c is plotted as a function of implantation cycle for the samples with MIMA technique in Fig. 4.9 [13]. Usually, the contact resistivity is limited on the range of 10^{-4} Ω cm^2 due to the low dopant concentration and severe Fermi level pinning effect [14]. The contact resistivity of MIMA sample decreases continuously with the increase of implantation cycle and a minimum value of 3.8×10^{-7} Ω cm^2 is obtained with four implantation cycles. The significant improvement in ρ_c is ascribed to the enhanced dopant activation in Ge by MIMA technique. Further increase in the implantation cycle may result in crystal damage due to the surplus implantation, which leads the ρ_c to rise again.

Figure 4.10 compares the ρ_c of metal/n-Ge contact achieved by different methods. Now, enhancing n-type dopant activation concentration and reducing electron SBH are the two main ways for obtaining low contact resistivity. For the n-type dopant activation enhancement method, a ρ_c of 6×10^{-5} Ω cm^2 is achieved by RTP annealing [15], and lower values of 7×10^{-7} and 8×10^{-7} Ω cm^2 are obtained through laser annealing and P/Sb co-implantation [16, 17], respectively. The lowest ρ_c of 3.8×10^{-7} Ω cm^2 is obtained by the MIMA technique. While for the electron SBH reduction method, the ρ_c values are reduced to 1.3×10^{-6} and 1.4×10^{-7} Ω cm^2

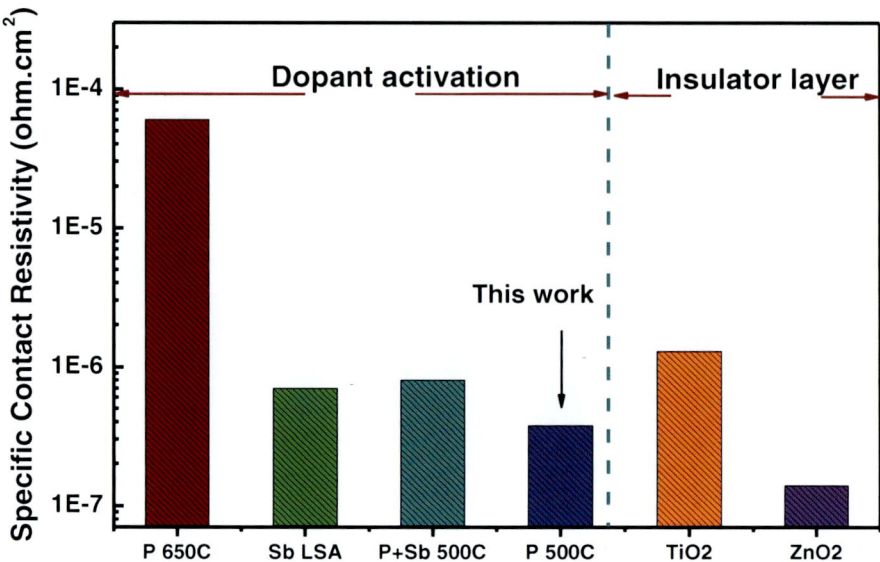

Fig. 4.10 Comparison of specific contact resistivity achieved by different methods

through inserting ZnO_2 and TiO_2 ultrathin layers [18, 19], respectively. It can be seen that the obtained ρ_c value is still larger than that of the ITRS target value (1×10^{-8} Ω cm^2) and the contact resistivity can be further reduced by combining dopant activation enhancement and electron SBH reduction methods.

4.2.4   Electrical Characteristics of Ge n⁺/p Junction

The impact of MIMA technique on the electrical characteristics of Ge n⁺/p junction Ge is also investigated in this thesis. The electrical characteristics of Ge n⁺/p diodes using MIMA technique in Fig. 4.11 exhibit a high on–off ratio over 10^5 with an ideality factor of 1.11, and the forward current in the series resistance limited regime is one order of magnitude higher compared to the single-implantation diode. The significant improvement of the forward current is ascribed to the enhanced n-type dopant activation in Ge, which is beneficial to reducing contact resistance between metal and n⁺-Ge. In order to evaluate the carrier conduction mechanism, the diode I–V characteristics are measured with different temperatures. By plotting $\ln\left(\frac{I}{T^2}\right)$ versus $\frac{1}{T}$ (Richardson plot) at different temperature (220–300 k), an E_a of 0.43 eV is obtained from the slopes of the Richardson plot. This value is close to the bandgap of Ge, verifying the diffusion-dominated leakage current and low defect density.

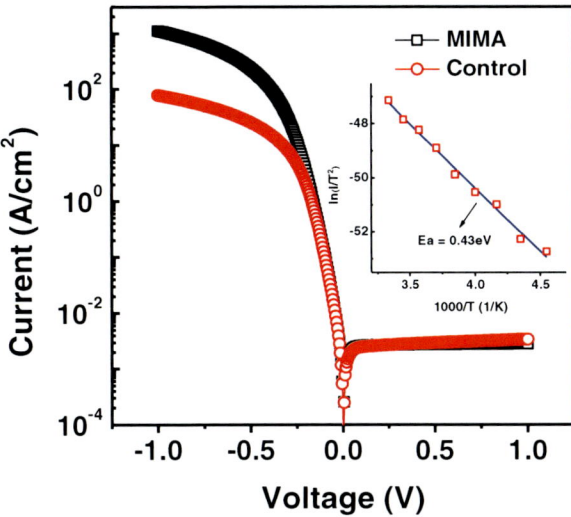

Fig. 4.11 The I–V characteristics of n⁺/p diodes with MIMA technique and single implantation measured at room temperature. The *inset* is temperature variation of leakage current density for n⁺/p diodes biased at −0.5 V

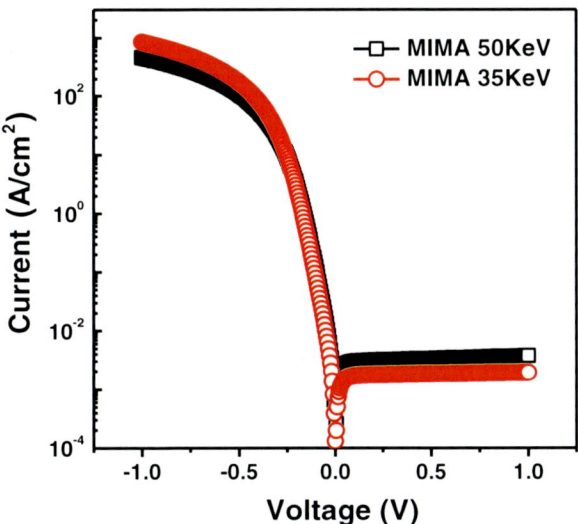

Fig. 4.12 The I–V characteristics of the MIMA n⁺/p diodes with different implantation energies

In order to reduce the junction depth, a lower implantation energy of 35 keV is adopted for the Ge n⁺/p junction fabrication with MIMA technique, and its electrical characteristics are given in Fig. 4.12. Compared to the Ge n⁺/p diode using 50 keV implantation energy, the Ge n+/p diode using 35 keV implantation energy exhibits an improved I–V characteristic, with a higher forward current and a lower reverse current. The improved electrical characteristics are ascribed to the reduced implantation damage with a lower implantation energy, which results in higher dopant activation concentration and lower defect density in Ge n+/p junction. Therefore, the junction depth can be reduced by maintaining good electrical characteristics by appropriately optimizing the implantation energy.

4.3   Ge-based nMOSFETs with MIMA Technique

As both the specific contact resistivity to n-Ge and electrical characteristic of Ge n+/p diode are significantly improved, the MIMA technique shows great potential to promote the performance of Ge nMOSFETs. Thus, the MIMA technique is integrated into Ge-based nMOSFETs in this section, and the corresponding device characteristics are also compared and analyzed.

4.3.1   Device Fabrication

Figure 4.13 gives the main process flow of Ge-based nMOSFETs. This fabrication step employed p-type (100) Ge wafers with resistivity of 1 ~ 10 Ω cm. After HCl rinsing and surface oxide removing, the Ge wafer was loaded into atomic layer deposition (ALD) equipment, and then nitrogen plasma treatment was carried out to passivate the Ge wafer surface. After that, 5 nm Al_2O_3 gate dielectric was deposited by ALD and 100 nm TiN was sputtered followed by inductively coupled plasma (ICP) etch to form a metal gate. After that, 20 nm SiO_2 was deposited for all the samples as capping layer to prevent dopant loss during activation. The P+ implantation at energy of 50 keV with a dose of 5×10^{14} cm^{-2} and rapid thermal annealing at 500 °C for 30 s in nitrogen ambient were in turn implemented four times. Finally, the capping layer was removed with diluted HF solution and the S/D contact holes were defined, Ti/Al metallization was performed to form the metal pads. The substrate contact was prepared directly through thermal evaporation of Al onto the Ge backside. Finally, the overall MOSFET structures were subjected to alloying annealing in N_2 ambient at a temperature of 375 °C for 30 min.

4.3.2   Device Characteristic Analysis

Figure 4.14 shows the I_d–V_g transfer characteristics of the Ge nMOSFET with MIMA technique. Relatively good transfer characteristics are observed with good I_{ON}/I_{OFF} ratio of ~10^2 and large I_{on} over 20 μA/μm, but with high leakage current over 10 nA/

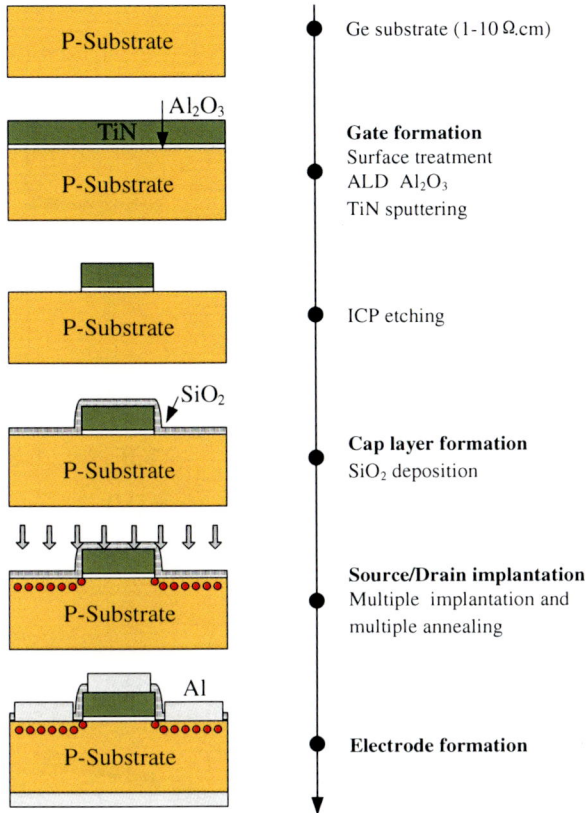

Fig. 4.13 The main process flow of Ge-based nMOSFETs

μm due to the small bandgap of Ge material. The I_d–V_{ds} output characteristics of Ge-based nMOSFETs with MIMA technique and single implantation are given in Figs. 4.15 and 4.16. The saturation current of Ge nMOSFETs with MIMA technique is 6.7 % larger than that of Ge nMOSFETs with single implantation, and this is due to the reduced parasitic resistance resulted by the enhanced n-type dopant activation concentration. In addition, the source/drain parasitic resistances of these devices are extracted in Fig. 4.17. The S/D parasitic resistance of the MIMA sample is 8.88 kΩ μm, which is smaller than that of single-implantation sample (19.74 kΩ μm), further verifying the reduced parasitic resistance by MIMA technique. It should be noted that the relatively small improvement in the saturation current is due to the relatively large channel length of these devices, in which the channel resistance is dominant in the total resistance. Therefore, the MIMA technique will further show its advantages in reducing S/D parasitic resistance with device dimension continuously scaling.

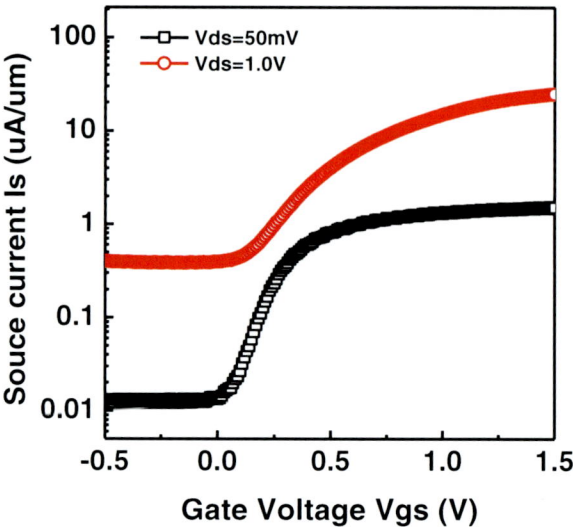

Fig. 4.14 Transfer characteristics of Ge nMOSFET with MIMA technique

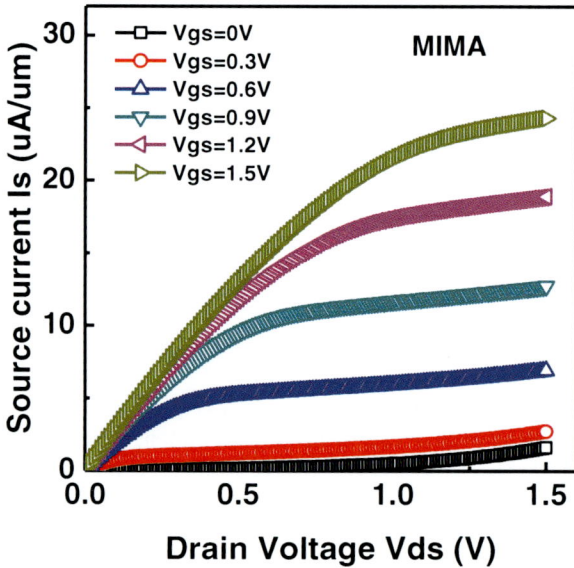

Fig. 4.15 The I_d–V_{ds} output characteristics of Ge nMOSFETs with MIMA technique

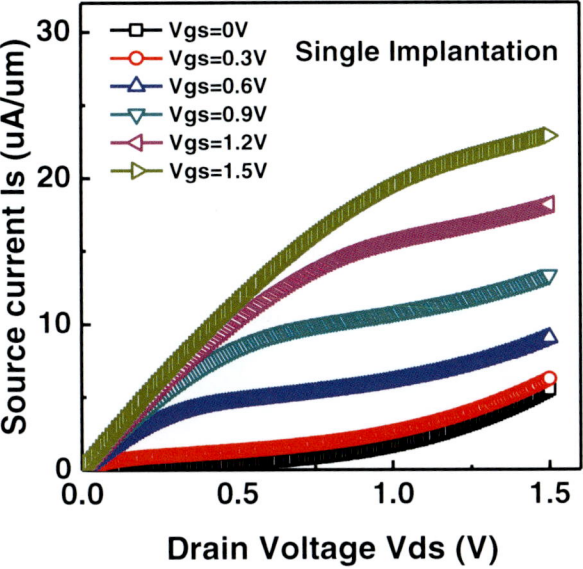

Fig. 4.16 The I_d–V_{ds} output characteristics of Ge nMOSFETs with single implantation

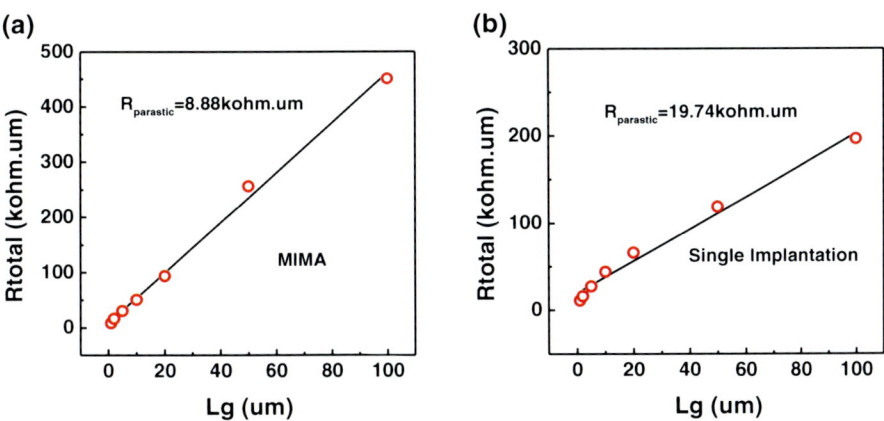

Fig. 4.17 The parasitic resistance extraction of **a** MIMA sample **b** single-implantation sample

4.4 Summary

With device dimension shrinking into nanoscale regime, devices suffer from a significant drive current reduction resulted by mobility degradation and S/D parasitic resistance increase. Although Ge-based MOSFETs improves its channel carrier mobility, the large parasitic resistance is a critical limiting factor for the performance improvement of Ge-based device. This chapter focuses on contact resistivity reduction of metal/n-Ge contact by MIMA technique, and the performance of Ge-based nMOSFETs is effectively improved by S/D parasitic resistance reduction.

(1) A high electrical concentration over 1×10^{20} cm^{-3} is achieved by the MIMA technique due to the obviously reduced implantation damages.
(2) The contact resistivity of metal/n-Ge contact is dramatically reduced to 3.8×10^{-7} Ω cm^2, which is very beneficial to S/D parasitic resistance.
(3) The electrical characteristic of Ge n$^+$/p junction is greatly improved by the MIMA technique, exhibiting a high on–off ratio over 10^5.
(4) The MIMA technique is successfully integrated into Ge-based nMOSFETs. Its saturation current is increased by 6.7 %, due to the reduced S/D parasitic resistance.

References

1. Noori AM, Balseanu M, Boelen P, Cockburn A, Demuynck S, Felch S et al (2008) Manufacturable processes for <=32-nm-node CMOS enhancement by synchronous optimization of strain-engineered channel and external parasitic resistances. IEEE Trans Electron Devices 55:1259–1264
2. International Technology Roadmap for Semiconductors (2013) Available http://www.public.itrs.net/
3. Kalra P (2008) Advanced Source/drain Technologies for Nanoscale CMOS. ProQuest, Ann Arbor
4. Kim SD, Park CM, Woo JCS (2002) Advanced model and analysis of series resistance for CMOS scaling into nanometer regime—part II: quantitative analysis. IEEE Trans Electron Devices 49:467–472
5. Kim SD, Park CM, Woo JCS (2002) Advanced model and analysis of series resistance for CMOS scaling into nanometer regime—part I: theoretical derivation. IEEE Trans Electron Devices 49:457–466
6. Oh JH, Chen Y-T, Ok I, Jeon K and Lee S-H (2010) High specific contact resistance of ohmic contacts to n-Ge source/drain and low transport characteristics of Ge nMOSFETs. In: International conference on solid state devices and materials, Japan, pp 3–20
7. Shayesteh M, Daunt CLM, O'Connell D, Djara V, White M, Long B et al (2011) NiGe contacts and junction architectures for P and As doped germanium devices. IEEE Trans Electron Devices 58:3801–3807
8. Chui CO, Gopalakrishnan K, Griffin PB, Plummer JD, Saraswat KC (2003) Activation and diffusion studies of ion-implanted p and n dopants in germanium. Appl Phys Lett 83:3275–3277
9. Brotzmann S, Bracht H (2008) Intrinsic and extrinsic diffusion of phosphorus, arsenic, and antimony in germanium. J Appl Phys 103:033508–033508-7
10. Kim J, Bedell SW, Sadana DK (2012) Multiple implantation and multiple annealing of phosphorus doped germanium to achieve n-type activation near the theoretical limit. Appl Phys Lett 101:112107
11. Othonos A, Christofides C, Boussey-Said J, Bisson M (1994) Raman spectroscopy and spreading resistance analysis of phosphorus implanted and annealed silicon. J Appl Phys 75:8032–8038
12. Schroder DK (2006) Semiconductor material and device characterization. John Wiley & Sons

13. Li Zhiqiang, An Xia, Yun Quanxin, Lin Meng, Li Min, Li Ming, Zhang Xing, Huang Ru (2013) Low specific contact resistivity to n-Ge and well-behaved Ge n$^+$/p diode achieved by multiple implantation and multiple annealing technique. IEEE Electron Device Lett 34(9):1097–1099
14. Martens K, Firrincieli A, Rooyackers R, Vincent B, Loo R, Locorotondo S, et al (2010) Record low contact resistivity to n-type Ge for CMOS and memory applications. In: 2010 International electron devices meeting—technical digest
15. Raghunathan S, Krishnamohan T, Saraswat KC (2010) Novel SiGe source/drain for reduced parasitic resistance in Ge NMOS. ECS Trans 33:871–876
16. Thareja G, Liang J, Chopra S, Adams B, Patil N, Cheng SL, et al (2010) High performance germanium N-MOSFET with antimony dopant activation beyond 1x10(20) cm(−3). 2010 International electron devices meeting—technical digest
17. Thareja G, Cheng SL, Kamins T, Saraswat K, Nishi Y (2011) Electrical characteristics of germanium n(+)/p junctions obtained using rapid thermal annealing of coimplanted P and Sb. IEEE Electron Device Lett 32:608–610
18. Lin J-Y, Roy AM, Saraswat KC (2012) Reduction in specific contact resistivity to Ge using interfacial layer. IEEE Electron Device Lett 33:1541–1543
19. Manik PP, Mishra RK, Kishore VP, Ray P, Nainani A, Huang Y-C et al (2012) Fermi-level unpinning and low resistivity in contacts to n-type Ge with a thin ZnO interfacial layer. Appl Phys Lett 101:182105

Chapter 5
Conclusions and Prospects

With aggressive downsizing of MOSFETs, its performance suffers from great challenges due to the mobility degradation and source/drain parasitic resistance increment. And therefore, new materials and novel processes will be necessary for performance enhancement in nanoscaled CMOS technology. Compared with other high mobility materials, such as III–V material, carbon nanotube, and graphene, Ge is an attractive material for nanoscale MOSFETs because of its superior properties, especially its higher and more symmetric carrier mobilities. Excellent performance has been demonstrated in Ge pMOSFETs, but some issues still need to be addressed for realizing high performance Ge nMOSFETs, and one of the challenges is the large parasitic resistance in S/D, due to the low solid solubility, high diffusion, and low activation of n-type dopant in Ge.

Metal S/D is a very promising approach for solving these mentioned problems in source/drain engineering. By adopting metal S/D to replace the traditional doped S/D, the limitations of the low dopant activation and low dopant solid solubility are overcome, but the performance of metal S/D MOSFET is still limited by other factors, including the high electron SBH and poor thermal stability of NiGe film. The high electron SBH results in obvious ambipolar behavior with insufficient drive current at the on-state and large leakage current at the off-state, and the poor thermal stability puts a restriction on device process window.

Therefore, this thesis focuses on the source/drain engineering of Ge nMOSFETs and several feasible solutions have been proposed. First, implantation after germanide (IAG) technique is proposed to modulate the electron SBH of metal/n-Ge contact, and extremely low electron SBH is achieved with optimized process parameters. Second, P and Sb co-implantation technique is adopted in this thesis to improve the thermal stability of NiGe, and the morphology and electrical performance of NiGe/Ge contact are great improved by this technique. Finally, low specific contact resistivity to n-Ge is achieved by Multiple Implantation and Multiple Annealing (MIMA) technique, which is very beneficial for the performance improvement of Ge nMOSFETs. The main work and conclusion of this thesis are given as follows:

© Springer-Verlag Berlin Heidelberg 2016
Z. Li, *The Source/Drain Engineering of Nanoscale Germanium-based MOS Devices*, Springer Theses, DOI 10.1007/978-3-662-49683-1_5

(1) Dopant segregation is adopted for modulating the electron SBH of NiGe/Ge contact. First, the ion implantation after germanidation (IAG) technique is investigated in detail with phosphorus (P) and arsenic (As) implantation, and the process parameters (including drive-in annealing and implantation condition) are also optimized. The current characteristics of NiGe/p-Ge diode changes from ohmic to well rectifying with I_{on}/I_{off} ratio over 10^5, and record-low electron SBH of 0.10 eV is achieved. Second, ion implantation before germanidation (IBG) technique is also studied with P and Se implantation, and the electron SBH is effectively reduced with exhibiting well ohmic characteristics of NiGe/n-Ge contact. Finally, the mechanism for SBH modulation is discussed, which is attributed to the formed dipoles at the NiGe/Ge interface resulted by dopant segregation.

(2) Co-implantation of P and Sb dopants into NiGe film is proposed to improve the morphology and electrical characteristics of NiGe/Ge contact. First, good morphology of NiGe is obtained with a reduced RMS roughness of 1.70 nm. Meanwhile, the thermal stability is enhanced to at least 600 °C, providing a wide temperature window for Ge-based MOSFET fabrication. Second, the electrical characteristics of the NiGe/Ge contact are improved by P and Sb co-implantation technique, behaving as a low leakage current of NiGe/p-Ge Schottky diodes and a good ohmic characteristic of NiGe/n-Ge contact with a small resistivity of 1.2×10^{-6} Ω cm^2. Finally, P and Sb co-implantation technique is successfully integrated into Ge-based SB nMOSFETs, and the device offers well-behaved output characteristics.

(3) The MIMA technique with P$^+$ implantation is applied to improve the contact resistance of metal on n-Ge. First, a high electrical concentration over 1×10^{20} cm^{-3} is achieved by MIMA technique due to the obviously reduced implantation damages. And the contact resistivity of metal/n-Ge contact is dramatically reduced to 3.8×10^{-7} Ω cm^2, which is very beneficial to S/D parasitic resistance. Second, the electrical characteristic of Ge n$^+$/p junction is greatly improved by MIMA technique, exhibiting a high on–off ratio over 10^5. Finally, the MIMA technique is successfully integrated into Ge-based nMOSFETs. Its saturation current is increased by 6.7 %, due to the reduced S/D parasitic resistance.

Based on current results, there are several avenues which can be pursued in future work.

(1) Now, the electron SBH is reduced to 0.1 eV by IAG technique, and a well-behaved ohmic characteristic is also realized in NiGe/n-Ge contact, but how to integrate the IAG technique into advanced device structures, such as FinFET, TFET and Nanowire, needs to be further designed and investigated.

(2) The thermal stability of NiGe film is enhanced to at least 600 °C, which meets the requirement of Ge-based device integration, but the thickness of NiGe film is relatively large, and improving the thermal stability of NiGe film for application in smaller dimension device is one future research direction.

(3) The contact resistivity achieved by MIMA technique ($3.8 \times 10^{-7} \, \Omega \, cm^2$) is still larger than that of the ITRS target value ($1 \times 10^{-8} \, \Omega \, cm^2$), and the contact resistivity may be further reduced by combining dopant activation enhancement and electron SBH reduction methods.

(4) The Ge n^+/p junction fabricated by MIMA technique shows well electrical characteristics, but its junction depth needs to be further controlled by adopting advanced annealing methods, such as flash annealing and laser annealing.

Finally, hoping the investigation in this thesis is helpful for the researcher in the relative research field.